AF577050

Forage Conservation Techniques

Forage Conservation Techniques

Pankaj Kumar Singh
Assistant Professor
Department of Animal Nutrition
Bihar Veterinary College
Patna-800014, Bihar

NEW INDIA PUBLISHING AGENCY
New Delhi – 110 034

NEW INDIA PUBLISHING AGENCY
101, Vikas Surya Plaza, CU Block, LSC Market
Pitam Pura, New Delhi 110 034, India
Phone: + 91 (11)27 34 17 17 Fax: + 91(11) 27 34 16 16
Email: info@nipabooks.com
Web: www.nipabooks.com

Feedback at feedbacks@nipabooks.com

ISBN: 978-93-83305-10-0

Composed, Designed and Printed in India

Preface

Sustainable livestock production depends mainly on feeding balanced and economical ration to livestock. Feed shortages both quantitatively and qualitatively limit livestock production. The cheapest source of high quality feed is forages. Forages are usually consumed fresh by domestic animals. However, it is possible to conserve them for use during future periods of feed shortages. When forage growth exceeds herd requirements, one of several strategies available to keep pasture quality high is fodder conservation as hay and silage making. Hay is preserved by drying. Hay is the oldest, and still the most important, conserved fodder, despite its dependence on suitable weather at harvest time. Silage involves natural fermentation, which produces lactic and other acids, which preserve the forage. Hay and silage helps to preserve forages at optimum nutritional value and ensure forage availability from the present to the future especially during drought, flood or disaster.

Forage Conservation Techniques was written in response to the need for a complete textbook on forage conservation methods and utility of the conserved forages to livestock. This book covers in-depth discussions of all theoretical and practical aspects of hay and silage preservation and their use in livestock feeding. Mechanism of silage and hay preservation, role of microbes and nutrients in forage preservation and quality assessment have been written in detail. A chapter on analytical techniques deals with methodology of analysis of nutrients and other parameters to judge the quality of hay and silages. Special emphasis has been given to recent advances in silage microbiology, preservatives and quality assessment, which will help to gain better understanding of ensiling process and to develop better additives. There is also a list of references at the end of each chapter, so that students can pursue the literature beyond that presented in this book.

The book is written in view of agricultural, biological, animal science and veterinary students as hay and silage preservation techniques are taught as a chapter or complete course in different under and post graduation studies. This book is essential reading for researchers, dairy development officers, feed manufacturers and livestock farmers.

I have consulted a number of books, journals and reports during the preparation of manuscript of this book. I wish my gratitude to all the writers and contributors of such literature. Special thanks to Dr. Chandramoni, Dr. Kashalendra and Dr. Sanjay Kumar, who encouraged me to write this book. I am indebted to my teachers, faculty members, students and friends for their help. I appreciate the contributions of my wife, Sarita Singh and my son, Arnav for providing me cheerful environment and patience throughout this endeavor. Above all I thank God and my parents for his guidance and inspiration.

I hope that this book will be a tool and an inspiration for future progress in the field of forage conservation technology.

July 2013 **Pankaj Kumar Singh**

Abbreviations

ADF	:	Acid Detergent Fibre
ADIN	:	Acid Detergent Insoluble Nitrogen
AIA	:	Acid Insoluble Ash
AOAC	:	Association of Official Analytical Chemists
BC	:	Buffering Capacity
cfu	:	Colony Forming Unit
cm	:	Centimeter
CF	:	Crude Fibre
CP	:	Crude Protein
CTAB	:	Cetyl Trimethyl Ammonium Bromide
DM	:	Dry Matter
EE	:	Ether Extract
FAO	:	Food and Agriculture Organization
FC	:	Fermentation Coefficient
FDA	:	Food and Drug Administration
h	:	Hectare
kg	:	Kilogram
km	:	Kilometer
LAB	:	Lactobacillus
mg	:	Miligram
N	:	Nitrogen
NDF	:	Neutral Detergent Fibre
NPN	:	Non Proteinous Nitrogen
ppb	:	Parts Per Billion
ppm	:	Parts Per Million
PVC	:	Polyvinyl Chloride
q	:	Quintal
VFA	:	Volatile Fatty Acid
W/W	:	Weight by Weight
WSC	:	Water Soluble Carbohydrate

Contents

CHAPTER 1

Importance of Forages in Animal Production

Livestock production is an integral part of agriculture and plays a significant economic role in most developing countries. It contributes to poverty alleviation, food security and provides elements that are essential to the national economy. Livestock production accounts for 40% of the gross value of the agricultural production globally and this figure is likely to go up, as the demand of livestock products is increasing rapidly with the increase in income and urbanization (FAO, 2012). For poor farmers, animal ownership ensures varying degrees of sustainable development and economic stability. Livestock provide a stable cash reserve independent of inflation and are an important source of traction. Livestock are closely linked to the social and cultural lives of several million people. Quality and quantity of feed is essential for optimum animal productivity. Inadequacy of quality animal feed resources is most often the crucial factor for livestock production. The success of livestock is greatly dependent on continuous supply of good quality nutritious feed at competitive price. Poor nutrition of animals has been identified as the major constraints to animal production across the developing world (FAO, 2000). Feed is the major input cost in animal production system, accounting for 65-70% of the total rearing cost.

Forages have always been an extremely important source of nutrients in ruminant's rations. The most important contribution of forages is their direct effect on livestock production (Mannetje, 1997). Forages represent a diverse range of feed stuffs that make a significant contribution to the overall nutritional economy of milk, meat and wool producing ruminants (Wilkins, 2000). Ruminants are used to convert forage into meat, milk and other products. Ruminant production systems are based on forages. They require forage as basal feed. Forage is defined as 'edible' parts of plants, other than separated grain, that can provide feed for grazing animals or that can be harvested for feeding. A wide range of feeds are included as 'forages' but they generally have substantial contents of crude fibre (more than 18%) and are suited to utilization by herbivores with their substantial capability for microbial digestion of the fibre. The 'forage crop' is often used to describe crops, generally annual or biennial, which are grown to be utilized by grazing or harvesting as a whole crop (e.g. maize, sorghum, berseem etc.). Forages include herbage, browse, straw, hay and silage (Barnes and Baylor, 1995). A wide range of crops in addition to perennial grass can be used as forage crops. Maize, oats, berseem, leucerne etc. are major forage crops used for ruminant feeding. Forage crops must have peculiar features like yield, seasonality of yield, quality etc. in order to warrant their inclusion in production system (Wheeler, 1986).

In most ruminant production systems, livestock derive between 40 and 90% of their feed requirements from forages. Forages provide more than 90% of the feed energy consumed by the herbivorous livestock of the world (Fitzhugh *et al.*, 1978). Grasses and other forages may provide nutrients at low cost, because of the high yields of dry matter and energy that can be obtained (Wilkins, 2000). Forages provide fiber in the ration which enhances proper digestion in forage-consuming animals. Ruminants have the ability to digest fiber via microbial fermentation and to obtain useful energy from this process. Ruminant's posses the enzymes capable of breaking down the complex α-linked polymers that form plant cell wall. In ruminants, rumen is the principal site of forage digestion, where the feed is retained for substantial period of time and subjected to extensive microbial fermentation under anaerobic conditions (Beever, 1993). Following hydration of the ingested feed, the microbial enzymes break down the complex dietary polysaccharides, namely cellulose and hemicelluloses, which are the principal component of the cell walls. The major end-products are the constituents sugars, principally glucose derived from cellulose, both hexoses and pentoses from hemicelluloses.

Monosacchardies derived from the dietary polysaccharides are converted to volatile fatty acids (acetic, propionic and butyric acid). Ultimately volatile fatty acids are metabolized to yield energy to ruminants (McDonald, 1999). Volatile fatty acids contribute approximately 70% of the total energy output.

Fibre in ruminant's rations is essential for animal health, since it is required to support an appropriate rumen function and physiology. Fiber in the forage also affects cud chewing, rumination and rumen health. An adequate amount of fiber is required in the diet to maintain normal animal health and rumen function. Ruminants require fibre in coarse physical form for a more effective chewing and ruminal activity. Increasing fibre content and forage particle size in diet effectively increases chewing activity resulting in increased saliva flow, rumen pH, acetate-to-propionate ratio, and milk fat levels. Forages are essential to maintain a metabolic balance in the rumen. The importance of the fiber content of forage increases as the level of milk production increases. Rumen acidosis often occurs when there is an insufficient amount of total fiber or effective fiber in the diet. Requirement of fibre in ruminants is usually met through green forages, preserved forages and agro industrial by-products. Producing high-quality forages can help to reduce the costs associated with feeding concentrates and supplements. Feeding natural grasses and green fodder will help to reduce the expenditure on concentrate feeding. Grasses are also good source of protein. One kg of green fodder gives 15 to 20 g protein to the animal. The cheapest source of high quality feed is well-managed pasture. Also, green grass is a good source of vitamin A which is present in the form of carotene. One kg of green grass provides 50 mg of vitamin A. This vitamin is necessary to maintain the health and reproductive status of the animal. With the appropriate grazing interval and intensity, pasture quality can be maintained at a high level throughout the year. Cutting fresh forage at the optimal stage of maturity and feeding it directly to animals year-round would supply the highest-quality and most palatable feed possible. The cost of producing livestock products has increased dramatically over the past few years as strong demand for feedstuffs has driven feed costs to record setting levels. Forages play a larger role in cattle production in response to the record-high grain and protein prices. Forages are a major economic source of nutrients in the ration of lactating dairy cows. Feeding average or poor-quality forages may limit milk production per cow and income-over- feed costs. There are several advantages of forage based feeding system as compared to concentrate based intensive one because forage based feeding system is cheaper than concentrate based feeding system thus, livestock can be raised with comparatively low input. Forage can improve the fatty acid profile of meat relative to a concentrate. Forage based feeding also maintains soil health and prevents soil erosion, thus protects environment (Peters, 2001).

However, fluctuations in seasonal growth and plant maturity make it necessary to harvest and store forages to maximize both quality and productivity. When pasture growth exceeds herd requirements in mid-spring, one of several strategies available to keep pasture quality high is fodder conservation. The aim of fodder conservation is about utilizing as much pasture as possible by conserving any that is a genuine surplus to animal requirements.

Why conserve forage?

Availability of fresh fodder varies throughout the year, but livestock must eat every day; fodder conservation is therefore desirable in most mixed farming systems. Provision of feed for deficit seasons has always been a major component in the improvement of animal production. When livestock are adequately fed in the lean season, their output increases. In any intensive ruminant production system, it is essential that the forage resource be stocked in such a manner so there is adequate forage quantity available per animal unit throughout the year. Silage and hay making is practiced widely in intensive animal production systems, mainly for two reasons. Firstly, because during the winter period there is no high quality feed available in the fields and secondly in order to feed high quality conserved supplements at any time of the year to complement grass to improve milk production and/or nitrogen utilization. Objectives of the silage making are:

- To preserve feed at optimum nutritional value
- To shift available feed from the present to the future
- To move feed from one location to another location
- To assist pasture management
- To fill feed gaps

- For production feeding
- For feeding management during scarcity like drought, flood or disaster.

Types of forage conservation

Forage, crop residues and by-products are usually consumed fresh by domestic animals. However, it is possible to conserve them for use during future periods of feed shortages. Nutritive value of conserved forages has a direct relation to animal production and land productivity (Bareeba et al., 1992). The conservation of forage and crop residues is a traditional way to reduce seasonal variations in availability.

The primary methods of preserving forage crops include silage making and hay making. Hay is preserved by drying and will generally keep while it is kept dry. Hay making involves reducing the moisture content of cut herbage in the field from an initial 70 - 90% to 15 - 20%. In drier climates, hay making is still important (Wilkinson et al., 1999). However, hay making is difficult in tropical regions because at the time when the forage is of acceptable quality (early in the spring season) to conserve it, the weather is likely to be too unreliable for sun drying. Hay is the oldest, and still the most important, conserved fodder, despite its dependence on suitable weather at harvest time. It can be made with simple equipment, manually or with mechanization, and many small-scale farming systems make hay to assure livestock feed through the lean season.

Silage involves natural fermentation, which produces lactic and other acids, which 'pickle' or preserve the forage. This fermentation takes place only under anaerobic (oxygen-free) conditions, so the forage must be packed to remove air and sealed to keep air out. Silage will generally keep while it remains sealed and anaerobic. It is recognized that the production of silage of high quality cultivated forage can be a valuable component for the development of a high-performing and low-cost system of animal production, using a relatively low level of purchased concentrates. Each of these methods of forage preservation has benefits and limitations that make it more desirable than the others for a specific livestock operation. However, forage conservation allows the intensification of dairy farming with a higher stoking rate than using nutrients from grasslands and harvested forages only.

References

Bareeba, F.B. 1992. Forage conservation and utilization. *Nomadic Peoples*. 31: 91-96.

Barnes, R, F. and Baylor, J.E. 1995. Forages in a changing world. In: Barnes, R.F., Miller, D.A. and Nelson, C.J. (eds.). Forages, volume. 1. *An Introduction of Grass land Agriculture,* Iowa State University Press, Ames, Iowa, pp.3-13.

Beever, D.E. 1993. Rumen function. In: Forbes, J.M. and France, J. (eds). *Quantitative Aspects of Ruminant Digestion and Metabolism*. CAB International, Wallingford. pp. 187-215.

FAO. 2000. Agriculture towards 2015/30. Technical Interim Report. April 2000. Economic and Social Department, Food and Agriculture Organization of the United Nations, Rome, Italy.

FAO. 2012. Crop residue based densified total mixed ration: A user friendly approach to utilize food crop by-products for ruminant production. T. K. Walli, M.R. Garg and H.P.S. Makkar (eds.). FAO Animal Production and Health. paper No. 172. Food and Agriculture Organization of the United Nations, Rome, Italy.

Mannetje, L. 1997. Potential and prospects of legume-based pastures in the tropics. *Tropical Grasslands* .No. 31, pp. 81–94

Peters, M., Home, P., Schimdt, A., Holmann, F., Kerridge, P.C., Tarawali, S.A., Schultze, R., Lascao, C.E.., Argel, P., Stur, W., Fujisaka, S, Muller-Samann, K. and Wortmannm S. 2001. The role of forages in reducing poverty and degradation of natural resources in tropical production systems . Network paper No.7. July , 2001. Agricultural Research and Extension Network. Centro Internacional de Agricultura Tropical (CIAT), A A/ 6713, Cali, Colombia,

Wheeler, J.L. 1986. Complementing grassland with forage crops. In: Morley, F.H.W. (ed.) *World Animal Science B. Disciplinary Approach,* I, Grazing Animals. Elsevier, Amsterdam. pp. 239-260.

Wilkins, R.J. 2000. Forages and their role in animal systems. In: Givens, D.I., Owen, E. Oxford, R.F.F. and Omed, H.M (eds.). *Forage Evaluation in Ruminant Nutrition*. CABI Publishing. CAB International. Wallingford, Oxon OX108DE, UK. pp. 1-14.

Wilkinson, J.M., Hill, J. and Leaver, J.D. 1999. Effect of swath treatment on water loss during field wilting and on feeding value of perennial ryegrass silage. *Grass Forage Science*. 54:227-236.

CHAPTER 2

History of Forage Conservation

Silage is the material produced by controlled fermentation of forages or crop residues with high moisture content. The objective of silage making is to preserve the harvested forages by anaerobic fermentation in which soluble carbohydrates are converted into acids. Due to acid produced, pH of a well-ensiled product becomes so low that all life processes come to a halt and the material remain preserved so long as it remains in airtight storage. Soluble carbohydrate present in forage is converted to lactic acid which drops the pH to a level sufficient to inhibit any further biological activity in the ensiled forage mass (McDonald, 1991). Fresh forage crops, such as maize, grasses, legumes, wheat and Lucerne, can be preserved by ensiling. It is essential to have a good microbial fermentation process to produce high quality silage. Good silage is achieved by discouraging the activity of plant enzymes and undesirable micro-organisms and encouraging the dominance of lactic acid bacteria. A good fermentation process is not only dependent on the type and quality of the forage crop, but also on the harvesting and ensiling technique. Silage-making is a managemental tool that allows producers to match feed resources (forages, crop residues, agro-industrial byproducts, etc.) with feed demand for a dairy herd. The basic function of silage-making is to store and preserve feed for later use with minimal loss of nutritional qualities.

History of silage making

Forages have been preserved using acids for many years and the process is referred to as ensilation. Silage making has its origins in antiquity. Ensiling is a natural process and was already used in ancient times by farmers around 3000 years ago (Kristein, 1963; Schukking, 1976). There is pictorial evidence on an Egyptian mural in the Museum at Naples dating from the period 1000 to 1500 BC, showing farmers filling a silo with sorghum-like crops. Kirstein (1963) recorded that silos were uncovered in the ruins of Carthage, an indicator that fodder was ensiled about 1200BC. Also the Roman historian and agriculture writer Caro (100AD) referred to the Teutons storing green fodder in the pits covered with dung. The importance of anaerobic conditions (airtight storage) and the danger of health entering a newly opened silo were already known in ancient times.

The first reference to the conservation of green fodder by ensilage was that of Professor John Symonds of the University of Cambridge in 1786, who during a tour of Italy observed the preservation of green leaves in wooden casks. The earliest account of the ensiling process as it understood today was given by Grieswald in 1842 (Watson and Smith, 1956) in which ensilage of fresh grass in pits was described. Attention was apparently paid to rapid filling of the silo, the exclusion of air, and provision of an effective seal from the silo. These factors are now accepted as essential prerequisites for successful ensilage. Silage making remained a techniques used in Mediterranean countries and it was not until the eighteenth century that silage making became common practice in other parts of Europe. The practice of simply throwing the fodder into excavated pit and covering it with earth, which the early farmers used, has been maintained in Europe to well into the twentieth century. Although grass was and still is the most common type of forage ensiled , maize was first ensiled in Hugary around 1860, and soon after in Germany.

An early book about silage making ("Manuel de la culture et de 1'ensilage du mais") was written by Goffart, a French farmer and writer, in 1877 and afterwards translated into Dutch and English. This book was also published in the USA, where it led to a huge development of growing and ensiling maize. Goffard advocated maize to be cut into 1 cm lengths, a practice that aided compaction of the material in the silage, which was beneficial for the exclusion of air, to be ensiled in brick silos.

Voelcker (1884) was probably the first person to draw the distinction between non-fermented and fermented silage which he described a" sweet silage" and sour silage, respectively. Sweet silage had undergone a lactic acid fermentation and "sour" silage an acetic or butyric fermentation, the former being greatly preferred because of the higher feeding value and absence of bacterial spores leading to spoilage of cheese. The prevention of butyric acid formation depends on the exclusion of air, not on allowing the grass to heat up. Whereas, the name referred to the sense of smell and not of taste. It was thought that the "warn" method was necessary to obtain "sweet", whereas cold method was necessary for "sour" silage. Sweet silage implied heat-cured forage in which sealing of the silo was delayed to permit the generation of heat. This heat, generated by plant respiration and aerobic microorganisms, can be sufficiently intense to cause the death of vegetable form of microorganisms. The conserved material then become cured and fails to ferment leading to formation of sweet silage. The method of production of sweet silage was known as "warm" method. The warm was developed by the English Scientist Fry (Fry, 1985) towards the end of the nineteenth century and further developed in Switzerland during the First World War, advised that forage, which was sometimes wilted to 30% dry matter, should be heated in the silo to at least 50°C, before the silo was closed airtight. However, this method led to large losses of dry matter and feeding value. Sour silage was formed from forage in which the silo was sealed from the air immediately upon completion of filling. Sour silage was developed by cold method. The "cold " method developed in Germany was based on preventing heating up of the material by making an air tight closure immediately after filling the silo.

Silage making was introduced into the United States by Professor F.H. King in 1981. He demonstrated the value of silage as an animal feed and its popularity in that country increased. In the Netherlands silage was first made of crop of spurrey (*Spergula sativa*) in 1845 on the Estate Hoekelum in Bennekom of Baron van Wassenaar in Bennekom of Baron van Wassenaar. The crop was harvested as it began to set seed and put into a water tight circular pit in layers of about 10 cm interspersed with thin layers of finely ground rock salt. It was reported that dairy cows relished the silage together with short chopped straw, despite the sour smell, resulting in well coloured butter.

A Danish agricultural extension officer (Pedersen) developed an artificial method of silage making during 1917 and 1918 by adding different acids (e.g hydrochloric acid) to freshly harvested and stored grass. However, around 1925 the Finish scientist A.I.Virtanen (AIV) achieved fame by advocating the practical application of acidification with hydrochloric acid and sulfuric acid (Virtanen, 1933). He added only as much acid as was necessary to lower the pH below 4. This way the losses due to respiration and protein breakdown were reduced leading to better product. However, the use of inorganic acids led to health problems in the animals, which developed "acid urine". Feeding of limestone was necessary to overcome this problem. Due to health problems the process began to lose its popularity in the early 1962s with the introduction of formic acid as a treatment as a direct acidification of silage crops. Ensiling is now a major conservation method for large-scale enterprises in sub-humid and humid climates, but it is laborious without heavy machinery.

History of Hay making

Storage of straws and crop residues for feed is probably as old as cereal cultivation. Storing forage for use in the non-growing season is an ancient practice which probably originated in countries where climatic conditions for hay making were good, where livestock were important and where herbage could dry off as standing hay. Before the agricultural revolution in western Europe, livestock were usually in very poor condition at the end of winter, and in some cases had to be helped to pasture when the vegetation began to grow again in spring. Stock not needed for reproduction or draught were slaughtered in autumn and the flesh salted. Winter feed was limited to some crop residues and, perhaps, a little meadow hay. The introduction of crops sown for hay (red clover, ryegrasses, lucerne) was a major step forward in improving winter feeding and crop rotation, with root crops (turnips, swedes, beet) complementing hay in some areas. In North Africa and western and central Asia, hay making, often from lucerne, has long been a very important source of off-season high-quality feed for both the sedentary and pastoral livestock sector. The domestication of lucerne was closely associated with the development of the horse as a war animal. The evolution of fodder growing and haymaking, until modern times, was more associated with draught animals, cavalry and transport than with dairying (FAO, 2000).

Soybeans have a long history as a nutritious hay and silage crop. Prior to World War II, the principal use of soybeans was as forage. By the early 1940s, soybeans were harvested at various stages of growth, cured and fed as high protein hay, which was considered to be similar in nutritional value to alfalfa (Blount *et al.*, 2002)

In the tropics, standing dried natural grass and the roughages derived from field crops are in most cases the cheapest feed for the dry season. Natural drying of mown herbage by sun and wind is still the commonest way of conserving fodder.

References

Blount, A. R. S., Wright, D. L Sprenkel, R. K., Hewitt, T. D. and Myer, R. O. 2002. Forage Soybeans for Grazing, Hay and Silage. Florida Cooperative Extension Service, Institute of Food and Agricultural Sciences, University of Florida. Publication No.SS-AGR-180

FAO. 2000. Hay and Straw Conservation - For Small-Scale Farming and Pastoral Conditions. Suttie, J.M. (eds.). FAO Plant Production and Protection Series No. 29. Food and Agriculture Organization of the United Nations. Rome, Italy.

Fry, G. 1885. *Sweet Ensilage*. Agricultural Press, London, pp.160

Kirstein, K. 1963. Historical survey of the ensiling of green fodder. *Das Wirtschaftseigene Futter*. 9:54-65.

Mannetje, L. 2000. Silage making in the tropics with particular reference on small holders. *Proceedings of the electronic conference on Silage making in Tropics*. September 1-November 30th 1999. FAO, Rome. pp. 161-180.

McDonald P., Henderson, A.R. and Heron, S.J.E. 1991. The Biochemistry of Silage. 2nd edition. Chalcombe Publications, Marlow, Bucks, UK.

Schukking, S. 1976. The History of Silage making. *Stikstof*. 19: 2-11.

Virtanen, A.I. 1933. The AIV method of processing fresh fodder. *Empire Journal of Experimental Agriculture*.1: 143-155.

Voelcker, A. 1884. The chemistry of ensilage. *Journal of the Royal Agricultural Society of England*. 20: 482-450.

Watson, S.J. and Smith, A.M. 1956. Silage. 2nd Edn, Crossby Lockwood.p.144.

CHAPTER 3

Silage Fermentation Processes

Ensiling is the term given for all physical and chemical changes that take place when forage with sufficient moisture content (60-65%) is stored in the absence of air for silage preservation. Silage fermentation is an exceptionally complex process involving biochemical interactions among the forage, microbial populations and the ensiling environment. Successful silage production depends upon the promotion of the fermentation brought by beneficial bacteria. The process is based on spontaneous lactic acid fermentation under anaerobic conditions. The lactic acid bacteria ferment the water-soluble carbohydrates in the crop to lactic acid, and to a lesser extent to acetic acid. Due to the production of these acids, the pH of the ensiled material decreases and spoilage micro-organisms are inhibited. The ideal characteristics of material for silage preservation are: DM content above 20% and an adequate level of fermentable substrate (8-10 percent of DM) in the form of water soluble carbohydrate. The ensiling material should also ideally have a physical form that allows easy compaction in the silo. Materials such as maize stover, sorghum stover and grass can be ensiled successfully, while crop residues such as rice and wheat straw, with low water soluble content, do not fulfill these requirements, and therefore pre-treatments, such as fine chopping or use of additives, or both, may be necessary.

From a practical view, the three most important things that must occur in order to make good silage are

1. The rapid removal of air,
2. The rapid production of lactic acid that results in a rapid drop in pH,
3. Continued exclusion of air from the silage mass during storage and feed out.

The Ensiling Process

Silage is essentially a battle between "good" and "bad" micro-organisms. The outcome of the battle, that is, whether a silage is well preserved or poorly preserved, depends not only on forage characteristics, but also in great part on the practices followed during the ensiling process (Catchpoole and Henzell, 1971) Once the fresh material has been harvested, chopped, compacted and sealed well to exclude air, the ensiling process then begins and undergoes, which can be described by a sequence of four phases as below (Weinberg and Muck, 1996; Merry *et al.*, 1997). These phases are:

Phase 1: Respiration, which degrades plant nutrients in the presence of oxygen

Phase 2: Early fermentation, which produces acetic acid, formic acid and other organic acids as a result of the growth of facultative aerobic bacteria such as enterobacteria, which can live in the presence (aerobe) or absence (anaerobe) of oxygen

Phase 3: Lactic acid fermentation by lactic acid bacteria that are strictly anaerobic, that is, they can only grow of oxygen

Phase 4: Stabilization phase due to the presence of lactic acid, which inhibits further degradation

Phase 1 - Aerobic phase

Until plant cells are dead or have no oxygen available, they respire to generate the energy they need for normal activity. As the forage is harvested, aerobic organisms predominate on the forage surface. During the initial ensiling process, the freshly cut plant material, and the aerobic and facultative aerobic micro-organisms such as yeasts and enterobacteria continue to respire within the silo structure. Plant respiration continues for several hours (and perhaps days if silage is poorly packed) until air is used up. Thus, any oxygen trapped between the forage particles is eliminated as a result of the respiration of plant material and the aerobic and facultative aerobic micro-organisms (yeasts and bacteria). If air is not removed quickly, high temperatures and prolonged heating are commonly observed. Air can be eliminated by wilting plant material to recommended dry matters for the specific crop and storage structure, chopping forage to a correct length, quick packing, good compacting, even distribution of forage in the storage structure, and immediately sealing the silo. Air must be removed before optimal fermentation can take place. Once air is removed, fermentation can begin. Lactic acid bacteria utilize water-soluble carbohydrates to produce lactic acid, the primary acid, responsible for decreasing the pH in silage. The acidity of silage can be determined by measuring its pH. Depending on the crop, plant material in the field can range from a pH of about 5 to 6 and decrease to a pH of 3.6 to 4.5 after acid is produced. In addition, a rapid decrease in pH will inhibit the growth of undesirable anaerobic microorganisms such as *enterobacteria* and *clostridia.* Another important chemical change that occurs during this early phase is the breakdown of plant proteins. Plant enzymes such as proteases and carbohydrases are active during this phase until air is used up, provided the pH is within the normal range for fresh forage juice (pH 6.5-6.0). Proteins are first reduced to amino acids and then to ammonia and amines. Up to 50 percent of the total plant protein may be broken down during this process. The extent of protein breakdown (proteolysis) is dependent on the rate of pH decline in the silage. A quick reduction in silage pH will help to limit the breakdown of protein in the silo by inactivating plant proteases. Phase I ends once the oxygen has been eliminated from the silage mass.

This phase is undesirable since the aerobic bacteria consume soluble carbohydrates that might otherwise be available for the beneficial lactic acid bacteria or the animal consuming the forage. Although, this phase reduces the oxygen to create the desired anaerobic conditions for silage fermentation but the respiration process produces water and heat in the silage mass. Excessive heat build-up resulting from an extended Phase I period can greatly reduce the digestibility of nutrients such as proteins. Under ideal crop and storage conditions, this phase will last only a few hours. With improper management, this phase could continue for several weeks. Delayed filling that result in excessive amounts of air trapped in the forage mass can have detrimental effects on the ensiling process. Rapid removal of air is important because it prevents the growth of unwanted aerobic bacteria, yeasts, and molds that compete with beneficial bacteria for substrate. Key management practices are proper maturity, moisture, chop length and rapid filling with adequate packing and proper sealing of the storage structure.

Practical aspect of the aerobic phase

- Chop the material as short as possible (1-3 cm)
- Fill the silo quickly
- Compact the silo as well as possible, as fingers should not be able to insert into the compacted forage.
- Seal the silo air tight
- Weight the top of the stack to maintain air tight seal between the cover and compacted forage
- Seal as soon as possible.

Phase 2 - Fermentation phase

After the oxygen in the ensiled forage has been utilized by the aerobic bacteria and the silage becomes anaerobic, Phase II begins. This is an anaerobic fermentation where the growth and development of acetic acid-producing bacteria occurs. These bacteria ferment soluble carbohydrates and produce acetic acid as an end product. Acetic acid production initiate the pH drop necessary to set up the following fermentation phases. As the pH of the ensiled mass falls below 5.0, the acetic bacteria decline in numbers as this pH level inhibits their growth. The lower pH enhances the growth and development of lactic acid producing bacteria. If the fermentation proceeds successfully, lactic acid bacteria develop and become the predominant population. The lactic-acid bacteria ferment soluble carbohydrates and produce lactic acid. Lactic acid is the most desirable fermentation acids and for efficient preservation, the acid should comprise greater than 60 percent of the total silage organic acids produced. When silage is consumed, lactic acid will also be utilized by cattle as an energy source. Due to the production of lactic and other acids, the pH decreases to 3.8-5.0. This low pH inhibits the growth of all bacteria. When this pH is reached, the forage is in a preserved state. No further destructive processes will occur as long as oxygen is kept away from the silage. Instead of lactic acid producing bacteria developing, large populations of clostridia bacteria may grow in the silage. These anaerobic bacteria produce butyric acid rather than lactic acid, which results in sour silage. With this type of fermentation, the pH may be 5.0 or above. The final pH of the ensiled forage depends largely on the type of forage being ensiled and the condition at the time of ensiling.

This is the longest phase in the ensiling process and continues for between several days to several weeks, depending on the properties of the ensiled forage crop.

Practical aspect of the aerobic phase:

- Mix molasses (@3-5% on wet basis), a substrate source for the bacteria, to encourage lactic acid fermentation.
- Maintain the dry matter about 30%.

Phase 3 - Stable phase

In general, good silage will remain stable once air is eliminated and it has achieved a low pH. For as long as air is prevented from entering the silo or container, relatively little changes occurs. Most micro-organisms of phase 2 slowly decrease in numbers. Some acid-tolerant micro-organisms survive this period in an almost inactive state; others, such as clostridia and bacilli, survive as spores. Only some acid-tolerant proteases and carbohydrases and some specialized micro-organisms, such as *Lactobacillus buchneri*, continue to be active at a low level.

Practical aspect of the aerobic phase:

- Maintain the airtight seal around the silage.
- Repair holes as soon as they are noticed.

Phase 4 - Feed-out phase or aerobic spoilage phase

This phase refers to the silage as it is being fed out from the storage structure in which surface of the silage is exposed to air. This phase is important because research shows that nearly 50 percent of the silage dry matter losses occur from secondary aerobic decomposition. High populations of yeast and mold or the mishandling of stressed crops can lead to significant losses due to aerobic deterioration of the silage. During feed-out this is unavoidable, but it can start earlier due to damage to the silage covering (e.g. by rodents or birds). The process of spoilage can be divided into two stages. The primary spoilage stage is the onset of deterioration due to the degradation of preserving organic acids by yeasts and, occasionally, by acetic acid bacteria. This will cause a rise in pH, and thus the second spoilage stage is started, which is associated with increasing temperature, and activity of spoilage micro-organisms such as moulds and enterobacteria. When exposed to air, spoilage micro-organisms metabolize lactic acid that causes the pH of the silage to increase, thus allowing bacteria that where inhibited by low pH to grow and further spoil the mass. However, the rate of spoilage is highly dependent on the numbers and activity of the spoilage organisms in the silage. Spoilage losses of 1.5-4.5% DM loss per day can be observed in affected areas. These losses are in the same range as losses that can occur in airtight silos during several months of storage (Honig and Woolford, 1980). Aerobic spoilage occurs in almost all silages that are opened and exposed to air. Airtight silos and removal of sufficient silage during feed-out can help to prevent aerobic spoilage.

To avoid failures, it is important to control and optimize each phase of the ensiling process. In phase I, good silo filling techniques will help to minimize the amount of oxygen present between the plant particles in the silo. Good harvesting techniques combined with good silo filling techniques will thus minimize water soluble carbohydrates losses through aerobic respiration in the field and in the silo, and in turn will leave more water soluble cabohydrates available for lactic acid fermentation in phase 2. During phases 2 and 3, the farmer cannot actively control the ensiling process. Methods to optimize phases 2 and 3 are therefore based on the use of silage additives applied at the time of ensiling, as will be discussed in the section on additives, below. Phase 4 will start as soon as oxygen is available. To minimize spoilage losses during storage, an airtight silo is required, and any damage to the silo covering should be repaired as soon as possible. During feed-out, spoilage by air ingress can be minimized by a sufficiently high feed-out rate. In addition, silage additives capable of decreasing spoilage losses can be applied at the time of ensiling. Proper management is vital to reduce these losses.

Practical aspect of the aerobic phase

- Maintain air tight seal.
- Feed out to ensure about 20-30 cm removal from the entire silage face each day.
- If the silage gets hot, feed it out at faster rate.

Phases of silage fermentation and storage

Phase	First	Second	Third	Fourth	Fifth
Age of Silage	***0-2 days***	***2-3 days***	***3-4 days***	***4-21 days***	***21 days-***
Activity	Cell respiration	Production of acetic acid, lactic acid and ethanol	Lactic acid bacteria	Lactic acid bacteria	Aerobic decomposition on re-exposure to oxygen
Temperature Change	69-90° F	90-84° F	84 ° F	84° F	84 ° F
pH Change	6.5-6.0	6.0-5.0	5.0-4.0	4.0	4.0-7.0
End Product	CO_2, heat and water	Acetic acid and lactic acid bacteria	Lactic acid	Lactic acid	Mold and yeast activity

Steps of silage formation

Although the ensiling process appears quite simple, many factors affect type and products of fermentation that takes place in a silo. Acid production is essential in the keeping qualities of silage. The rate, amount and kind of acid produced are influenced by the moisture content of the chopped forage and the readily available carbohydrate content of the forage.

1. Harvest forage at proper stage of maturity

The crop for silage is harvested at proper stage of maturity, which ensures good quality silage. The crop is generally harvested at the flowering stage when it has the maximum amount of nutrients. Maize is harvested at the early dent stage whereas sorghum is cut at late dough stage.

2. Chop to proper length

Prior to ensiling, plant materials should be chopped. Chopping facilitates facilitate good compaction while filling. A good compaction results in a good anaerobic fermentation and prevents fungal growth and rotting of the silage. When most oxygen is removed, clostridial growth is discouraged and lactic acid fermentation encouraged. Secondly, chopping releases plant juices, stimulating the growth of lactic acid bacteria. Also, chopping may increase silage intake by improving quality of fermentation and by accelerating rate of passage of feed particles through the rumen. Chopping forages too long makes compaction difficult and air will remain trapped in the silage resulting in heating and spoilage, whereas, chopping the forage to cut any finer can have a negative impact upon milk fat production and the incidence of displaced abomasums in dairy cattle due to inadequate scratch factor. Thus, 10-15 percent of the silage material should be above 25 mm in length in order to maintain an effective fibre function.

Usually forage is chopped into short lengths (1-3 cm) before ensiling. But, the fineness of chopping varies with moisture content and nature of the material. Rough and hard materials should be finely chopped, while delicate and soft materials can be roughly chopped. The following guidelines can be used:

High moisture forage (Moisture >75%)	chop to 6.5-25 mm
Wilted materials (Moisture 60-70%)	chop to 6.5 mm
Whole maize plant	chop to 6.5-13 mm

3. *Control of moisture content in raw materials*

Ensilage can only be successful only when the moisture content of the raw material is kept to a suitable level. Adequate moisture is essential for bacterial fermentation and aids in packing to help exclude oxygen from the silage. Although silage may be made within a large range of moisture contents, DM should be over 35-40 percent to assure good quality silage. Forages with DM content above 50% are considered difficult to ensile (Staudacher *et. al.*, 1999). The dry matter content of the forage can also have major effects on the ensiling process via a number of different mechanisms. First, drier silages do not pack well and thus it is difficult to exclude all of the air from the forage mass. Second, as the dry matter content increases, growth of lactic acid bacteria is curtailed and the rate and extent of fermentation is reduced. (For example, acidification occurs at a slower rate and the amount of total acid produced is less).

There are many disadvantages to ensiling crops with high moisture content. Ensiling of wet materials results in the generation of a large volume of effluent not only poses disposal problems, but also carries off valuable, highly digestible nutrients in solution. Effluent is produced when moisture is above 75 percent. Also, ensiling wet crops will encourage clostridial fermentation, resulting in excessive protein degradation, high DM loss, reduced nutritive value and production of toxins. Even if the water-soluble carbohydrate levels are adequate to ensure lactic fermentation, very wet silages may still be nutritionally undesirable because voluntary dry matter intake of these is frequently low. Where weather permits, wilting forage above 30-35% DM prior to ensiling can reduce the incidence of clostridia because these organisms are not very osmo-tolerant (they do not thrive well in dry conditions).

Maturity also has a tremendous impact on moisture of forage crops. Moisture content of forage and grasses is usually above 70 percent when harvested at a suitable stage. Therefore, the moisture content should be reduced by field wilting. When weather conditions are unfavourable, field wilting should be avoided to prevent nutrient loss due to rain leaching. In contrast, the moisture content of cereal straw is generally too low to allow tight packing, so cereal straw and stover should be finely chopped. Sometimes water should be added to bring moisture content to a suitable level. Sweet potato and agro-industrial byproducts such as pomace are high in moisture content and wilting is necessary. Chopped vines are usually mixed with finely chopped straw or bran prior to ensiling to increase overall dry matte content.

The moisture content of plant materials may be measured with instruments, but it is usually estimated manually on farm. Samples of chopped and minced grass or leguminous forage are grasped tightly by hand for one minute or so to estimate moisture content. If juice can be extracted, moisture content is above 75 percent. If the material remains together but without juice, moisture content is between 70-75 percent. If the material has elasticity and spreads out slowly, moisture content is 55-65 percent. If the material spreads out quickly, moisture content is about 55 percent. If the material breaks, moisture content may be below 55 percent.

4. Control of water soluble carbohydrates

Another factor that can affect the ensiling process is the amount of water-soluble carbohydrates present for good fermentation to take place. The amount of water-soluble carbohydrates necessary to obtain sufficient fermentation depends on the dry matter content and the buffer capacity of the crop. The buffering content of the forage mass can have an effect on silage fermentation. Alfalfa has a high buffering capacity in comparison to maize. Thus, it takes more acid production to lower the pH in alfalfa than in maize silage, resulting in the former being more difficult to make. Maize silage harvested at the proper growth stage or at physiological maturity has a high level of readily available carbohydrates for lactic acid production. If legumes and grass crops are not wilted in the field to an average of 65 percent moisture or less depending on the type of storage, then the addition of a carbohydrate-rich feedstuff will enhance fermentation.

Forages with insufficient fermentable substrate or too low a DM content, addition of sugars directly (e.g. molasses) or by adding enzymes that release extra sugars from the crop ensures sufficient substrate for lactic acid bacteria thus causing the desired reduction in pH.

5. *Filling, packing and sealing*

The silo should be filled with the crop harvested as rapidly as possible. Filling delays will result in excessive respiration and increased silage losses. Remove all air from the silo by compacting the fodder by jumping and stamping on the fodder after addition of each load. Any air left in the silo or any air entering the silo will render the microenvironment within the silo conducive for undesirable fungal and bacterial growth that will decompose the silage. Packing should begin immediately when storing silage in bunker silos. A wheeled tractor is preferred as a packing vehicle, as it will supply greater weight per surface unit than a tracked vehicle. The silo should be sealed with an air-tight cover once it is filled to prevent penetration of air and rainfall into the silage. A good grade of plastic weighted down with discarded tires will generally provide an adequate seal. Cover the silo and make it airtight by adding heavy materials like stones on the top and waterproof by providing a roof. The silage is generally ready after 4-6 weeks. Maintain airtight seal until feeding out. It has been observed that water soluble carbohydrate dramatically decreased and dry matter losses increased when maize forage was not immediately packed into silos after chopping. Losses increased with prolonged times of delay.

6. *Additives*

Most silage additives are designed to aid fermentation by providing fermentation bacteria, enzymes or fermentable substrate. Although not a replacement for good management, they are tools to help ensure that the ensiling process stays within acceptable boundaries. The types and numbers of bacteria on the plant also have profound effects on silage fermentation. Natural populations of lactic acid bacteria on plant material are often low in number. In addition, if air is not removed from the silage mass, other types of fermentation can occur. Proper maturity assures adequate fermentable sugars for silage bacteria and maximum nutritional value for livestock. In high DM silages with reduced water availability, the presence of suitable, osmo-tolerant lactic acid bacteria could become a limiting factor in the ensiling process. It has been shown that these bacteria represent only a small percentage of the indigenous microflora on forage crops (Pahlow and Weissbach, 1996). Inoculants that increase lactic acid fermentation might be useful to inhibit clostridial activity. The minimum number of lactic acid bacteria required to inhibit clostridial activity was found to be at least 100 000 colony-forming units per gram of fresh crop (Kaiser and Weiss, 1997).

7. *Feed to cattle*

If the cattle refrain from eating it while introduced for the first time, let the silage rest in open air for a while to get rid of some of the heavy odour gases, and then feed it to the cattle.

References

Catchpoole, V.R. and Henzell, E.F. 1971. Silage and silage-making from tropical herbage species. *Herbage Abstracts*. 41:213-21.

Honig, H., and. Woolford, M K .1980. Changes in silage on exposure to air. In: C. Thomas (ed.) Forage Conservation in the 80s. *Occasional Symposium No. 11*. British Grassland Society, Hurley, Berkshire, UK. pp 76-87.

Kaiser, E. and Weiss, K. 1997. Fermentation process during the ensiling of green forage low in nitrate. 2. Fermentation process after supplementation of nitrate, nitrite, lactic-acid bacteria and formic acid. *Arch. Anim. Nutr*. 50:187-200.

Merry, R.J., Lowes, K.F and Winters, A. 1997. Current and future approaches to biocontrol in silage. In: V. Jambor, L. Kalapil, P.Chromec, and P. Prochazka. (ed.) 1997. *Proc. 8th Int. Symposium Forage Conservation*, Brno, Czech Republic.Research Institute of Animal Nutrition, Pohorelice, Czech Republic pp17-27.

Pahlow, G., and Weissbach, F. 1996. Effect of numbers of epiphytic lactic acid bacteria (LAB) and of inoculation on the rate of pH-decline in direct cut and wilted grass silages. In: D.I.H. Jones, R. Jones, R. Dewhurst, R. Merry, and P.M. Haigh (ed.) *Proc. 11th Int. Silage Conf.*, Aberystwyth, UK. 8-11 Sept. 1996. IGER, Aberystwyth, UK. p.104-105.

Staudacher, W., Pahlow, G and Honig, H. 1999. Certification of silage additives in Germany by DLG. In: T. Pauly (ed.) *Proc. 12th Int. Silage Conference*, Uppsala, Sweden, 5-7 July. Swedish University of Agricultural Sciences, Uppsala, Sweden. pp 239-240.

Weinberg, Z.G and Muck, R.E. 1996. New trends and opportunities in the development and use of inoculants for silage. *FEMS Microbiol. Rev.* 19: 53-68.

CHAPTER 4

Silo

A silo is an airtight to semi-airtight structure designed for the purpose of preservation and storage of high moisture feeds as silage.

Role of the silo in silage making

Silo has only three functions in the ensiling process:

1. To provide a solid surface to permit compaction of the mass to eliminate air;
2. To protect the ensiled materials from air and water during the storage period;
3. Where desired, to aid in the removal of silage by providing a base for unloading equipment.

If the silo meets the first two requirements, the cost of its construction is of little significance. Thus, a trench silo dug in firm soil is as useful as a specially engineered and highly automated upright. The role of the silo in mechanization of silage handling should not be confused with its role in silage fermentation. For the preparation of good silage, the silos have the following characteristics:

1. ***Dimension of the silo:*** The size of the silo should be decided on the basis of the number and kind of animals to be fed daily, the length of feeding period, and the amount of forages available for ensiling. It should be of adequate depth for better packing and less surface area to total mass exposed. If the silo is below

the ground, the depth of the silo will depends on the water table in the soil. It should always be above the water table. In case of tower silo, the height of the silo will depend upon the machinery available for filling the tower silos. In silos there should be adequate provisions for the escape of surplus juices, either by drain or by a gravel bottom.

2. ***Air tight walls:*** The silos should have air-tight walls without any cracks whether they are below or above the ground. If they are above the ground the air may enter through the cracks and moulds would grow. If the silo is below the ground the rain water may enter and spoil the silage.

3. ***Smooth walls of the silos:*** The walls of the silo should be smooth and strong. The wall should be strong and rigid in order to withstand the pressure which develops inside the silo as fermentation takes place. Care should be taken to avoid the corners on constructions. The *kaccha* silo should be plastered with clay so as to prevent the seepage of effluent produced in the silage making.

4. ***Location of the silo:*** Silos should be conveniently located and accessible in all kinds of weather, from the standpoint of both filling and feeding. The silo pits (not tower type silo) should always be located preferably at the highest spot on the farm to avoid water seepage.

Types of silos: Silos are of different types

I. Upright or Tower silo

Tower silos are circular in shape and constructed from brick. Tower silos are usually circular in shape and equipped with a series of doors about 2 feet square approximately every 6 feet up one side of the silo. The size varies from about 12-20 feet in diameter and 40-80 feet. in length depending upon the requirement. The advantages of this type of silo include: durability, small space required, minimum top and side spoilage losses, and possibility for mechanization. However, tower silos are expensive due to its cost of construction.

II. Pit silo

A pit silo is like a tower silo but inverted into the ground. It resembles like a well. This type of silo can be made only in places where water table level is low enough that the silo will not fill with water. It is the most common silo on individual farms and suitable for rural conditions. Size can be adjusted according to scale of production. Round or square concrete silos are usually built inside houses for protection from the weather. Advantages are lower cost and easy management. Pit silos are never damaged by storm. A disadvantage pit silo is high effluent loss, especially with clay walls.

III. Trench silo

Trench is a horizontal silo generally built underground or semi-underground, with two solid walls of 1.5-2 m in height. It is most popular in areas where weather is not too severe and where there is good drainage. Low initial cost, ease of construction and more suitability for mechanization are advantages of trench silo. The tractor can be driven on top from one side to the other for compaction purposes. After compaction, it is sealed with a plastic sheet pressed down with soil, sand bags or straw bales to maintain anaerobic conditions. Trench silo require low initial cost of construction as compared to tower silo however it require larger space o seal.

IV. Bunker silo

Bunker silos are silos above the ground usually with concrete floor. Such silo saves labour.

V. Stack silo

This type of silo implies a pile of material on the ground surface. On flat and dry ground, plastic sheet is placed underneath and the material is laid in a stack. The top is covered with plastic and sealed all round with soil. Sandbags or old types, or any other suitable objects, are placed on top to prevent the top cover from being blown away by the wind. The advantages of the stack silo are low cost and flexibility of placement.

VI. Plastic silo

The plastic silo is similar to the stack silo but it is covered with plastic sheets of polyvinyl chloride (PVC) or polyethylene. Alternately, the silo can be made in bags with sealed tops. The stack silo is also inexpensive and can be placed anywhere. However, labour requirements are high due to manual filling and handling.

VII. Plastic bags

Forage ensiled inside small bags should be stored inside a second bag as the thin plastic is easily punctured. To ensure a tight seal, the neck should be twisted and tied or taped, then doubled over and retied or retaped.

Advantages of plastic bags

Plastics silage bags are an economical alternative to traditional silage storage systems, such as pits and silos when related, harvest and storage losses are considered.

- It is an effective way for preserving feed with minimum nutrient loss. (The anaerobic environment that is created eliminates spoilage from the growth of yeasts, molds and adverse bacteria while maintaining essential proteins and nutrients).
- Allows farmers to store silage anywhere they need it. A well graded and well drained ground surface is all that is necessary.
- The silage is completely sealed in the bag. This means that all the acid is retained in the silage unlike that in pit silage when it seeps out through the bottom of the pit as effluent. This compensates for the longer pieces of forage and poorer compaction than that found with silage machinery, so that the quality of the silage is just as good.
- Ensiling in a bag avoids the hard work of having to remove silage, as it has to be from a pit, when it has to be dug out every day.
- Because the whole bag is fed out to the animal, it means the rest of the silage which is in the other bags is not exposed to air at removal and is therefore unspoiled. Much of the silage in pits has been found to be spoiled due to poor sealing and exposure to air every day when the silage is removed for feeding.
- The bag is easily stored and easily portable so that any member of the family can carry it to the feed trough for the cow.

Management of silo

Silo dimension

The first step in forage storage design is to determine the volume of silage to be stored. Quantities should be determined on a DM basis to avoid the weight and volume discrepancies associated with As-

Fed (AF) moistures. To make the calculation, it is needed to know how much of each forage will be represented in the ration, again on a DM vs. AF basis. Once the volume of storage is known, the second step is to determine silo dimensions. To do so, a daily face removal rate needs to be selected. Face removal should be determined based on the minimum amount required to prevent heating of the face between feedings. A suggested minimal face removal rate of 6" per day for hay silage and 4" per day for maize silage (Bodman and Holmes, 1997) with the larger of the two doubled for design purposes. A width of at least 16 feet is needed to facilitate packing. To increase labor efficiency, design for a bottom width of 30 feet or more. This helps in positioning the transport vehicle close to the silo and minimizes loader turn around time. To calculate the silo width that allows the removal of one foot of silage face per day, it is necessary to determine the total weight of silage to be fed each day. Silo width may be adjusted by selecting a different wall height. Now that the silo width is known, length can be determined by:

Silo Length = Face Removal (inches/day) x Storage Period (days)

Inventory Management

Prior to selecting the number of structures to build, it is advisable to review aspects of both filling and inventory management.

The recommended procedure for filling a bunker silo or silage pile is to spread the silage in thin layers on the sloped filling face and driving over it several times with one or more heavy tractors. This is the "progressive wedge" method of silo filling. The top of the forage mass is a large area exposed to air and moisture penetration. Migration of air and moisture supports aerobic microorganisms, which cause deterioration throughout the top surface of exposed forage. The progressive wedge technique of filling continually covers previous layers of forage, thus reducing exposure to the elements. Packing the forage to form a high-density mass reduces entrapped air and limits penetration of both air and moisture into the forage. Filling the structure quickly (within three days) limits forage exposure to air and precipitation.

Consequently, equipment to harvest, transport and fill the storage as well as labor should be capable of filling rapidly. Each of two smaller bunkers/piles can be filled and covered with less exposure time than one large structure (Holmes and Muck, 2000). Additionally, management of varying qualities of forages into and out of one long structure is next to impossible. Multiple smaller silos can be filled more quickly, can contain forage of similar quality, exposes less surface area to the environment and allows the delivery of forage to different silos simultaneously for more efficient and complete packing under custom harvest conditions. When sizing storage, the selection of multiple smaller structures allows many management options over one large silo. One practical approach may be to size one or two silos for maize silage, one or two silos for first crop alfalfa and one silo for both second and third crop alfalfa. Locate silos for convenient year round access. A north-south orientation results in the fewest problems with snow accumulation (Bodman and Holmes, 1997).

Silo Design and Surface Area Exposure

It is advisable to reduce surface area exposure. The shallower the silage mass, the more top surface is exposed to the elements and the greater the DM loss. In small silos it may be necessary to reduce the sidewall depth or reduce the feeding rate to keep width at least 16 feet. If design of the silo precludes a silo width of at least 16 feet and wall heights of 8 feet, consider using alternative storage options like silo bags or a well-constructed pile. The importance of minimizing the amount of surface area of exposed silage was given by Bolsen *et al*, (1993). It was documented an increased loss of $1,720 per year for a silo with 8- foot versus 12-foot walls due to the increased surface area of the top when the feed is valued at $100/T DM. Lowering the height of the walls necessitates increasing the width of the silo from 50 feet to 75 feet to maintain storage capacity.

The preferred material for silo sidewalls is concrete. Because of contamination concerns, do not use pressure treated wood for feed storage. Other materials commonly used for temporary storage walls, including bales of forage or concrete "spoil blocks" should be avoided (unless lined with plastic) because of air penetration leading to higher rates of spoilage (Bodman and Holmes, 1997). Silo bags are used to form walls for bunker silos. Although used with some success, the level of management needs to be high to avoid unnecessary spoilage resulting from bag damage during feedout. When emergency storage is needed, it is recommended to use either silage bags or well packed silage piles.

Silo floors should be sloped 0.5 to 1.0% (6 to 12 inches per 100 feet) towards the open end. This slope is sufficient to drain off precipitation and melted snow and still allow safe operation of equipment across the slope. Except in rare instances, a concrete floor is a must. The concrete should be air-entrained concrete with gravel aggregate, formulated for a minimum compressive strength of 4,500 psi and have a minimum thickness of 6 inches. Asphalt has been used for silo floors and costs approximately one half to two thirds that of concrete. However, it requires a well-designed base, is more susceptible to heaving and equipment damage and is usually of shorter life than that of good concrete.

Filling Procedure

It is necessary to have proper silo filling to improve silage quality. To increase the packing factor or the DM density proper filling of silo is required. There are several procedures used for filling bunker/pile silos that are influenced by wall configuration and storage size. The filling procedure often begins by filling the back end of the storage by pushing forage up a sloped filling face in a progressive wedge technique. In the preferred method of filling, the forage is added in thin (<6") layers to this filling face until the storage is full. This is known as the progressive wedge method. The progressive wedge method allows a plastic cover to be applied to the top surface soon after that area is full. The progressive wedge method can be used with or without a back silo wall. When a back silo wall is built into an embankment or a ramp is constructed to the top of the back wall, forage can be deposited onto the filling face as the transport vehicle unloads while driving on the forage. Unloading while driving on the forage increases forage distribution efficiency and adds some packing of the forage. A blade/bucket is used to spread the forage in thin layers. Do not lift dump bodies on the forage as differential settling may cause a dump truck to tip over. These trucks are usually emptied on solid surfaces with the forage pushed into the bunker/pile with the tractor. Where bunker silo side walls are embanked and a 12 feet drive is left between bunkers, some producers use side discharge wagons to deliver forage along the wall by driving past the wall. This technique is most compatible with a horizontal fill procedure. The horizontal fill procedure places forage in the storage in horizontal layers. The horizontal fill procedure is not recommended for several reasons.

1. A much larger surface is exposed to air and precipitation during filling compared to the progressive wedge method where covering occurs during filling.

2. A storage which is not filled with one type and/or quality of feed will have multiple layers of forage. This eliminates the ability to inventory feed and blend rations according to ration need.
3. The whole cover must be removed to add more forage to fill the storage.

When filling silage piles, thin layers of forage must be placed on the filling face and driven upon to provide good packing and high density. This requires shallow slopes so a tractor can safely pack the whole surface. Driving on the forage in perpendicular directions will probably provide a denser mass. It is safer to back up a sloped surface than to drive forward or along the slope. Take care to avoid a tractor overturn. The shallow slope also assures plastic can be held tightly to the surface after covering.

Filling of silo

It is necessary to have rapid harvest and silo filling to improve silage quality. To initiate lactic acid fermentation as quickly as possible, the silo must be filled as fast as possible. Filling the silo evenly and properly compacting the silage crop are important to expelling atmospheric oxygen from the silo.

The silo filling speed should not be determined by the forage harvester, but by the compacting tractor. The packing tractor should be as heavy as possible to achieve adequate to high packing density. If the roller tractor is relatively light, it should be fitted with narrow tires and drive slowly to increase the pressure per surface unit and extend the time during which the pressure is exerted. The dimensions of the roller tractor should always be adjusted to the ensiling chain to ensure smooth and trouble-free operations and to avoid bottlenecks. The manufactured tractor weight can be augmented by adding weight to the tractor. Weight can be increased by adding iron wheel weights, adding liquid to tires, or adding front end and 3-point hitch weight. Dual wheels all around will improve traction and tractor maneuverability on a slippery surface. This added safety factor as well as the ability to add more tractor weight makes this a good choice.

Packing Density

Forage should be packed tightly to obtain high density. A high density in a silo is important for two primary reasons. Firstly and most importantly, density and dry matter content determine the porosity of the silage. Porosity, in turn, sets the rate at which air moves into the silo and subsequently the amount of spoilage which occurs during storage and feedout. Secondly, the higher the density, the greater the capacity of the silo. Thus, higher densities generally reduce the annual cost of storage per ton of crop by both increasing the amount of crop entering the silo and reducing crop losses during storage.

The factors affecting density in bunker and pile silos are not well understood. General recommendations have been to spread the crop in 6-inch layers and pack continuously with heavy, single-wheeled tractors (Holmes and Muck, 1999b).

Covering

Once filling is complete, immediate covering with an air-excluding material is required. The silo must be sealed airtight by covering it with suitable silage wrap which is weighted down by spreading tyres, sand bags or earth over its entire surface. Avoid damaging the wrap since this would allow air to enter. Plastic films have proven effective sealers when properly installed. Weighting the plastic cover with soil and tires has been effective at limiting air movement under the bunker silo plastic. Spreading soil on the perimeter edges of plastic covered silage piles helps to seal those edges and to keep the plastic from blowing in the wind. Plastic bag ends must be tightly sealed. Any holes in plastic covers should be repaired immediately after they occur.

Storage

The fermentation process requires about three weeks. The storage/ feeding process can last 50 weeks or longer. During the storage/ feeding period, the opportunity for feed value loss can be great. Loss of dry matter occurs primarily as aerobic (presence of oxygen) microbial deterioration of silage. If oxygen is allowed to contact the silage, the microbes use the oxygen and the silage (energy source) to grow and multiply. Oxygen can penetrate sound concrete but moves through cracks much more quickly. Sealing concrete with an epoxy or plastic sheets helps to reduce the amount of oxygen entering the silage through silo walls. Holes in plastic covers and bags should also be repaired to exclude air. If the cover on a silo does not preclude water, rainfall and runoff carry dissolved oxygen into the silage mass. This oxygen is also available to microbes for aerobic silage deterioration. Percolating water washes away organic acids resulting in an elevated pH. Along with the dissolved oxygen, this provides a good environment for the microbial population to grow, especially in warmer weather. The gaseous losses of are largely due to aerobic deterioration. Percolating water exiting the silo as effluent carries nutrients meant for the animals, thus increasing dry matter loss. This is another potential source of water quality degradation. Thus any cover used on a silage surface should be placed in such a way that water sheds away from the silage rather than onto or through it. Shape the top of the silage in a bunker to shed water from the silage after forage settling has occurred. Exclude runoff from flowing between the silage and bunker walls. This can be done by forming flow channels several feet from the walls and sloping toward the back of the bunker. Water flows over the plastic and exits the bunker onto the ground.

Bolsen *et al.* (1993) studied the effect of cover type and time of application on a 3.5-foot deep bunker silo. This study also considered the effect of exposure time in storage. It was observed that covering immediately with a plastic cover results in dry matter recovery exceeding 85% (15% DM loss) at all depths for storage periods up to 180 days. The top 13 inches experienced the largest dry matter loss. Depths greater than 13 inches had losses in the range of 5-8%. The average dry matter loss of 9% compares very closely with the 9.4% (corrected for seepage) found by Minson and Lancaster (1965) in their plastic covered bunkers. When no cover was used, significant dry matter loss occurred to the 26-inch depth. Loss values of 62% in the top 13 inches and 34% in the next 13 inches were experienced after a 180-day storage period. This 34% average loss compares closely to the 32% (corrected for seepage) found by Minson and Lancaster (1965) in their uncovered bunker silos. By delaying covering for a period of 7 days, Bolsen *et al.* (1993) was able to demonstrate significant dry matter losses in the top 13 inches during the delay period. After the cover was added, further dry matter loss was similar to that of the immediately covered bunker. Thus covering is effective on a "better late than never" basis but is most effective when applied immediately. This data and analysis strongly support the recommendation of covering the bunker silo/pile immediately with a material which excludes oxygen and rain water.

Feed Out

Silage feeding from storage should not occur until the fermentation process is complete. The full fermentation phase takes between two and three weeks to stabilize the silage. Fully fermented forages resist these losses of dry matter through the production of acid that lowers the pH to 4.0 or less. At that time, fermentation sugars are depleted and bacteria die off leaving a stable storage environment. If feed is required at an earlier time, a second, smaller silo should be set up. Never open silos whose fermentation process is still ongoing. Opening the silage mass prior to the stable phase of fermentation results in dairy cow digestive upsets. Once feedout occurs, the silage is exposed to oxygen, allowing yeast and molds to convert nutrients into carbon dioxide, water and heat. Feedout losses can represent up to 30% of the total dry matter loss in the ensiling process (Roth and Undersander, 1995). Density and feed out rate affect the amount of air exposure prior to feeding. In loosely packed silage, oxygen may move several yards into the silage from the open face. In contrast, densely packed silage limits the rate of oxygen diffusion into the feed out face. Practices and systems that promote high silage density should be used during filling to help minimize losses at feed out (Holmes and Muck, 1999a).

The feed out rate (inches removed from the silage face per day) can influence the average loss during feed out. The feed out rate (inches removed from the silage face per day) can influence the average loss during feed out. The rate and method of silage removal from the face critically affects feedout loss and animal performance. Losses occur from the exposed face and top as well as from loose silage lying on the floor between feedings. Therefore, only the amount of silage that will be fed in a short period should be uncovered at one time. Plastic can be pulled back from the silage top or cut off each day. At no time should more than three days worth of silage be exposed otherwise spoilage happens. Particularly top spoilage even when covered. Spoilage also occurs from the spillage of loose material during feedout and from exposed areas as a result of damage to the cover (Muck and Holmes, 2000).

If most dry matter loss occurs within 1-2 inches of a silage feed out face and the feed out rate is one inch per day, then all of the silage fed will have had some loss before being fed. Let's say as an example, 10% of that one inch of material was lost. Then all feed removed will have a 10% loss. On the other hand, if the removal rate is 4 inches per day and only the first inch experiences the 10% dry matter loss and the other 3 inches are unaffected, the average dry matter loss for the feed removed is 2.5%. Thus, it is important to size and manage the feed out rate for adequate rates of removal to minimize spoilage losses at the feed out face. Pitt and Muck (1993) determined the dry matter loss during feed out of bunker silos as a function of silage removal rate. They determined the dry matter loss was 3% at the recommended removal rate of 6 inches per day for 35% dry matter silage with a density of 14 lbs DM/ft^3. They also concluded dry matter loss was reduced as silage density increased. Muck and Pitt (1994) stated that dry matter loss is proportional to silage porosity. Porosity is inversely related to dry matter density and dry matter content Therefore, removal rates should never be lower than 4 inches per day in the summer and 3 inches in the winter. Minimal removal rates are most critical with hay crop silages, high moisture maize and drier silages (Bodman and Holmes, 1997).

One method of determining if enough silage is removed at each feeding is to take a sample 12 inch boring into the removal face at the beginning of a feeding and check for heating. If adequate silage is being removed and the appropriate removal management is employed so a tightly packed, smooth face results, the sample should be cool.

The problem of aerobic deterioration after opening the silo

If the silo is opened after the appropriate fermentation time, good silage will be sufficiently stable to withstand a couple of hot summer days without warming up to a significant degree. Silages whose acidity has been lowered fast, and to which no or only a simple biological silage additive has been added, may suffer from a shortage of stabilizing acetic acid. Although such silages have fermented well, they will warm up quickly after an influx of air into the silo. Specifically if there are still pockets of air in the overall silo (through faulty spreading and compaction) the air will affect the silage and harmful yeasts will multiply. They breathe the lactic acids in the silo, causing the pH value in the silo to increase and mould spores to sprout. Decay will start and if this is combined with a low feed-out speed and continuously high temperatures, major losses, forage waste and a poor feed quality cannot be avoided. Silage batches exposed to these negative influences are dangerous constituents for total mixed ration, since they will immediately contaminate the rest of the feed mixture with yeasts and moulds.

References

Bodman, G.R. and Holmes. B.J. 1997. Managing and Designing Bunker and Trench Silos (AED-43). Mid West Plan Service, Ames, IA.

Bolsen, K.K., Dickerson, J.T., Brent, B.E., Sonon, R.N., Dalke, B.S. Jr., Lin, C. and. Boyer, J.E Jr. 1993. Rate and extent of top spoilage in horizontal silos. *J. Dairy Sci.* 76:2940.

Holmes, B.J. and Muck, R.E. 1999a. Factors Affecting Bunker Silo Density. http://www.uwex.edu/ ces/crops/uwforage/storage.htm.

Holmes, B.J. and Muck, R.E. 1999b. Bunker Silo Packing Density Calculator. http://www.uwex. edu/ces/crops/uwforage/storage.htm.

Holmes, B.J. and Muck, R.E. 2000. Preventing Silage Storage Losses. University of Wisconsin- Madison. http://www.uwex.edu/ces/crops/uwforage/prevent-silage-storage7.PDF

Minson, D.J. and Lancaster, R.J. 1965. The Efficiency of Six Methods of Covering Silage. *New Zealand J. Agric. Res.* 8:542-554.

Muck, R.E. and Pitt, R.E. 1994. Aerobic deterioration in corn silage relative to the silo face. *Transactions of the ASAE.* 37(3):735-743.

Pitt, R.E. and Muck, R.E. 1993. A diffusion model of aerobic deterioration at the exposed face of bunker silos. *Journal of Agricultural Engineering Research.* 55:11-26

Roth, G. and Undersander, D. 1995. Corn Silage Production, Management, and Feeding. North Central Regional Publication No.574.

CHAPTER 5

Silage Microbiology

Successful preservation of high-moisture forage and other crops depends upon the controlling the activities of microbes, particularly bacteria. Ensiling generally controls microbial activity by a combination of an anaerobic environment and a natural fermentation of sugars by lactic acid bacteria on the crop. The silage microflora plays a key role in the successful outcome of the conservation process. Therefore, it is important to understand both the microorganisms that are present on the crop at ensiling and how ensiling preserves the crop, inhibiting detrimental microorganisms. The crop at ensiling contains both aerobic and anaerobic microorganisms and a range of both bacteria and fungi that affect silage quality.

The flora can basically be divided into two groups, namely the desirable and the undesirable micro-organisms. The desirable micro-organisms are lactic acid bacteria. The undesirable ones are the organisms that can cause anaerobic spoilage (e.g. clostridia and enterobacteria) or aerobic spoilage (e.g. yeasts, bacilli, listeria and moulds). Many of these spoilage organisms not only decrease the feed value of the silage, but also have a detrimental effect on animal health or milk quality, or both (e.g. Listeria, clostridia, moulds and bacilli).

A. Desirable microorganisms

1. *Lactic acid bacteria (LAB)*

The microorganisms which are most important to the ensiling process are the lactic acid bacteria (LAB). The term, lactic acid bacteria, covers bacteria from a number of genera (Lactobacillus, Pediococcus, Lactococcus, Enterococcus, Streptococcus and Leuconostoc) that are found in silage (Pahlow *et al.*, 2003). They primarily ferment sugars to lactic acid, but they also produce some acetic acid, ethanol, carbon dioxide and other minor products. All lactic acid bacteria that are regularly associated with silage are facultative aerobes (able to grown in the presence and absence of oxygen) and mesophilic (i.e. they can grow at temperatures between 5° and 50°C, with an optimum between 25° and 40°C). Lactic acid bacteria are normally present on growing crops in small numbers but usually multiply rapidly after harvesting, particularly if the crop is chopped or lacerated. The lactic acid bacteria generally need various amino acids and vitamins for growth (Pahlow *et al.*, 2003). Yet in spite of their complex requirements, they dominate the fermentation of crops in the silo once anaerobic conditions are established. When the crop is ensiled the lactic acid bacteria continue to increase and ferment the water soluble carbohydrate in the forage (Holzapfel and Schillinger, 1992; Hammes *et al.*, 1992; Teuber *et al.*, 1992; Weiss, 1992). The LAB comprises a rather large group of bacteria, encompassing species belonging to six genera. Based on their sugar metabolism lactic acid bacteria can be classified as obligate homofermenters (*Pediococcus damnosus* and *Lactobacillus ruminis)*, facultative heterofermenters (*Lactobacillus plantarum*, *Lactobacillus pentosus*, *Pediococcus acidilactici*, *Pediococcus pentosaceus* and *Enterococcus faecium*) or obligate heterofermenters (*Leuconostoc*, *Lactobacillus brevis* and *Lactobacillus buchneri*. Obligate homofermenters produce more than 85% lactic acid from hexoses (C_6 sugars) such as glucose, but cannot degrade pentoses (C_5 sugars) such as xylose. Facultative heterofermenters also produce mainly lactic acid from hexoses, but in addition they also at least degrade some pentoses to lactic acid, and acetic acid and/or ethanol. Obligate heterofermenters degrade both hexoses and pentoses, but unlike homofermenters they degrade hexoses to equimolar mounts of lactic acid, CO_2 and acetic acid and/or ethanol (Schleifer and Ludwig, 1995). The homofermentative groups are more efficient in converting forage sugars into lactic acid. Of these species *Lactobacillus plantarum* is generally considered to be the most competitive, rapidly producing large amounts of lactic acid in freshly ensiled forage. As a result of lactic acid production, pH of the silage reduced to 4 and 5, depending on the species and the type of forage crop. Homofermentative fermentation, which is the more desirable, yields lactic acid, which is a stronger acid than acetic acid, and does not produce carbon dioxide, which represents a loss of DM. Thus, homofermentative fermentation results in little or no loss of DM and only a small energy loss. However, even a totally heterofermentative process will rarely result in more than a 5% loss of DM, and the gross energy loss is negligible

Lactic acid bacteria are essentially non-proteolytic organisms and are therefore considered to be beneficial in that they assist in conserving labile proteins and free amino acids in forage.

B. Undesirable microorganisms

1. *Clostridia*

Clostridia are a group of undesirable bacteria that grow in silage, especially in moist silage (below 30% DM) under anaerobic conditions. Clostridia are obligate anaerobes and endospore-forming bacteria, which enables them to survive for a long time, even under bad conditions. Their effects on silage quality usually occur long after the lactic acid bacteria have stopped actively growing in the silo. Many clostridia ferment carbohydrates as well as proteins, thus causing problems such as the reduction in feeding value and the production of biogenic amines. Soil and manure are the most common sources of clostridia, therefore great care should be taken to avoid contamination of the forage with soil (e.g., from a soiled tractor wheel) or manure during harvesting. When silage is contaminated with dirt or manure, the risk of clostridial fermentation increases.

Clostridia can be divided into two major groups, saccharolytic clostridia and proteolytic clostridia. The saccharolytic clostridia (*Clostridium butyricum* and *Clostridium tyrobutyricum*) can degrade lactic acid and residual water soluble carbohydrates to butyric acid, H_2 and CO_2, causing a rise in pH (Gibson, 1965; Klijn *et al.*, 1995).

$$2 \text{ lactic acid} \rightarrow 1 \text{ butyric acid} + 2\ H_2 + 2\ CO_2$$

The second group comprises proteolytic species that ferment amino acids to a variety of products. The proteolytic group includes *Clostridium perfringens*, *Clostridium bifermentans* and *Clostridium sporogenes*. Essentially, clostridia dominate the fermentation when lactic acid bacteria do not produce enough lactic acid to drop pH to a stabilization value fast enough. Clostridia tend to growth faster at a temperature of about 35°C. Thus this type of undesirable fermentation happens when extensive respiration and enterobacterial fermentation occur and silage temperatures rise in the early phases of the fermentation process. These bacteria ferment amino acids to a number of different products including acetic acid, ammonia, amines and volatile fatty acids. Many clostridia ferment carbohydrates as well as proteins, thus causing problems such as reduction in feeding value and the production of biogenic amines. Biogenic amines have a negative effect on silage palatability (Van and Dulphy, 1996).

Clostridial activity is undesirable for several reasons. First and foremost is the reduced intake by livestock of silages. Amines have been associated with reduced intake in small ruminants but not necessarily in cattle. Also, the fermentation of lactic acid to butyric acid is one of the most wasteful anaerobic fermentations in silage, producing a 51% loss of DM and 18% loss of gross energy (McDonald *et al.*, 1991). Production of carbon dioxide and hydrogen gas represent losses of digestible DM and energy, and production of ammonia is a loss of protein. Also, butyric acid is a weaker acid than LA, and clostridia produce only one molecule of butyric acid from two molecules of LA or sugar. Such fermentation reduces the DM content by about 50%, and the energy content by almost 20%; it is referred to as secondary fermentation, and is accompanied by increasing pH. Some species of clostridia ferment sugars and change the lactic acid produced by lactic acid bacteria into butyric acid, carbon dioxide and hydrogen (H_2). Production of carbon dioxide and hydrogen gas indicates a loss of digestible energy. The breakdown of lactic acid into butyric acid, which is a weaker acid, means that the pH of silage going through clostridial fermentation will tend to rise. Butyric acid has a strong, repulsive smell. Trace amounts of butyric acid suffice to decrease voluntary intake by cows. The weakness of the acid and the production of ammonia both raise the pH, and high levels of ammonia also depress the animals' intake. Silage spoiled by clostridia is easily recognizable due to its strong odor, a pH above 5.0, ammonia nitrogen greater than 10% of total nitrogen higher amine content, and more butyric acid than lactic acid (McPherson and Violante, 1966). Ensiling methods that cause a rapid and sufficient drop in silage pH will help to prevent the development of such clostridial silage, because, clostridia are inhibited at low pH. Furthermore, clostridia are more susceptible to a low availability of water than lactic acid bacteria (Kleter *et al.*,

1982). For this reason, decreasing the water content of a crop, such as by wilting to a higher DM content, can be a way of selectively inhibiting clostridia (Wieringa, 1958). Finally, clostridia will also be inhibited by nitrite and NO, or compounds that are degraded in silage to nitrite and NO (Spoelstra, 1983, 1985). Some clostridia can cause serious health problems. One extremely toxic species is *Clostridium botulinum*. This organism can cause botulism (Kehler and Scholz, 1996), which can be deadly for cattle. Fortunately, *Clostridium botulinum* has a limited acid tolerance, and does not grow in well-fermented silage. Furthermore, clostridial spores present in silage are transferred to milk via faeces (clostridial spores can survive the passage through the alimentary tract of a dairy cow) and faecal contamination of the udder impair milk quality. Clostridial fermentation is characterized by a foul smell, indicative of the production of components such as putrescine and cadaverine; its many other negative effects include poor preservation of forage and silage, high pH, high DM losses, high ammonia-N production, low intake, and butyric acid production.

Thus, it is important to prevent clostridial bacterial activity. Fortunately, many clostridia are more sensitive to high acidity and high osmotic pressure (high levels of dry matter content) than lactic acid bacteria. Thus clostridial fermentation can be avoided by:

- Ensiling at more than 30% dry matter content;
- Packing the silo as densely as possible to reduce increases in temperature due to respiration;
- Avoiding soil contamination;
- Ensiling forages with the highest possible sugar content
- Using proper ensiling techniques that favor maximum fermentation to the lowest possible pH as soon as possible.

2. *Enterobacteria*

Enterobacteria (*Escherichia coli and Erwinia herbicola*) associated with silage are facultative anaerobes and usually are active in the early stage of fermentation when the pH (7.0) is favourable for their growth. The enterobacteria, sometimes described as 'acetic acid bacteria' or 'coliform bacteria', are usually present in very low numbers on crop. They compete with the lactic acid bacteria for the water soluble carbohydrates and ferment them to mixture of products including acetic acid, ethanol and hydrogen. In addition they can degrade protein, decarboxylate and deaminate amino acids leading to the production of large concentration of ammonia. This protein degradation not only causes a reduction in feeding value, but also leads to the production of toxic compounds such as biogenic amines and branched chair fatty acids. Biogenic amines are known to have a negative effect on silage palatability (Woolford, 1984; McDonald *et al.*, 1991; Van and Dulphy, 1996), especially in animals that are not yet accustomed to the taste (van *et al.*, 1997). Moreover, the ammonia formed through proteolysis increases the buffer capacity of the ensiled crop, thus counteracting any rapid decrease in silage pH. A special characteristic of enterobacteria is their capability to reduce nitrate (NO_3) to nitrite (NO_2) under silage conditions. In silage, nitrite can be degraded by enterobacteria to ammonia and nitrous oxide (N_2O), but it can also be chemically degraded to NO and nitrate (Spoelstra, 1987). With air, NO is oxidized into a mixture of gaseous, yellow-brown nitrogen oxides (NO_2, N_2O_3, N_2O_4). Gaseous NO and NO_2 have a damaging effect on lung tissue and can cause a disease with pneumonia-like symptoms known as "silo filler's disease" (Woolford, 1984). To prevent animals from being in contact with gaseous nitrogen oxides, they should not be housed in buildings adjoining silos during silo filling or the first week of silage storage (O'Kiely *et al.*, 1999). As a consequence, their fermentation is less desirable than that of lactic acid bacteria.

Despite the above-mentioned problems, a little nitrite reduction is considered positive for silage quality, because the nitrite and NO formed are very effective inhibitors of clostridia (Woods *et al.*, 1981; Spoelstra, 1985). Enterobacteria will not proliferate at low pH. Ensiling methods that induce a rapid and sufficient drop in silage pH will therefore help to decrease enterobacterial growth (McDonald *et al.*, 1991).

3. *Bacilli*

Bacilli are endospore-forming, rod-shaped facultative aerobes bacteria, Facultative aerobic bacilli ferment a wide range of carbohydrates to compounds such as organic acids (e.g. acetate, lactate and butyrate) or ethanol, 2,3-butanediol and glycerol (Shlegel,1987). Proliferation of bacilli in silage is generally considered undesirable. Not only are bacilli less efficient lactic and acetic acid producers than lactic acid bacteria (McDonald *et al.*, 1991), they can also enhance aerobic deterioration in later stages of ensiling (Lindgren *et al.*, 1985). However, their activity under anaerobic conditions is considered to be of relatively rare importance. Their more significant activity is in advancing the spoilage of silage when exposed to oxygen. After yeasts or acetic acid bacteria have raised pH (approximately 4.5 or above), and temperature has increased modestly (40°C), a second wave of heating usually occurs with temperatures rising to 50°C or higher. This often is the activity of bacilli (Muck & Pitt, 1994), producing a silage that has a slimy feel.

Furthermore, high numbers of *Bacillus* spores in raw milk have been associated with high spore numbers in fresh cow faeces (Giffel *et al.*, 1995). It seems very plausible that bacillus spores are transferred from silage to milk via feces similar to clostridial spores. Psychrotrophic *Bacillus cereus* spores are considered to be the most important spoilage organism of pasteurized milk (Giffel, 1997). High numbers of these (psychrotrophic) *Bacillus cereus* spores have been found in silages (Giffel *et al.*, 1995).To decrease bacillus growth in silage, storage temperatures should not be too high (Gibson *et al.*, 1958). Some specific *Bacillus* spp. is able to produce antifungal substances, and have been used to inhibit aerobic spoilage of silage (Phillip and Fellner, 1992; Moran *et al.*, 1993).

4. Acetic acid bacteria

Acetic acid bacteria are obligate aerobic bacteria that are capable of growing at low pH. They grow on ethanol, producing acetic acid. However, once ethanol has been exhausted, they can grow on acetic acid, producing carbon dioxide and water. This will raise pH and permit other aerobic microorganisms to grow. Consequently, acetic acid bacteria can be initiators of aerobic deterioration. The acetic acid bacteria that have been isolated from silage belonging to the genus *Acetobacter* (Spoelstra *et. al.*, 1988). The activity of Acetobacter sps. in silage is undesirable because they can initiate aerobic deterioration, due to the fact that they are able to oxidize lactate and acetate to carbon dioxide and water. Generally, yeasts are the main initiators of aerobic spoilage, and acetic acid bacteria are absent, or play only a minor role. However, for whole crop maize silages there is evidence that acetic acid bacteria alone can initiate aerobic deterioration (Spoelstra *et al.* 1988). Furthermore, selective inhibition of yeast also can increase proliferation of acetic acid bacteria in silage (Driehuis and van Wikselaar, 1996).

5. ***Listeria***

Members of the genus *Listeria* are aerobes or facultative anaerobes. Regarding silage quality, the most important species is the facultative anaerobe *Listeria monocytogenes,* because this species is pathogenic to various animals and man. Silage contaminated with *Listeria monocytogenes* has been associated with fatal cases of listeriosis in sheep and goats (Wiedmann *et al.,* 1994). Poor quality silage have identified as one of the main sources of contamination of raw milk by *Listeria monocytogenes* (Sanaa *et. al.,* 1993). Growth and survival of *Listeria* spp. in silage are determined by the degree of anaerobiosis, and the silage pH. *Listeria monocytogenes* can tolerate a low pH of 3.8-4.2 for long periods if oxygen is present, even if only at low levels. Under strictly anaerobic conditions, it is rapidly killed at low pH. Silages that have a higher chance of aerobic surface spoilage, such as big bale silages, seem to be particular liable to *Listeria* contamination (Fenlon *et al.,* 1989). *Listeria monocytogenes* generally does not develop in well-fermented silages with a low pH. So far, the most effective method to prevent growth of *Listeria monocytogenes* is to keep the silage anaerobic (McDonald *et al.,* 1991).

6. ***Yeasts***

Yeasts are eukaryotic, facultative anaerobic, heterotrophic micro-organisms. Yeasts are perhaps the most significant aerobic microorganisms on the crop relative to silage quality. Yeasts grow on soluble substrates, sugars and lactic acid being the most important relative to silage. In most circumstances, yeasts are the first group of microorganisms to develop once oxygen comes in contact with silage, either during storage or during feed out. In silages, anaerobic as well as aerobic yeast activity is considered undesirable. Under anaerobic silage conditions, yeasts ferment sugars to ethanol and CO_2 (Schlegel, 1987; McDonald *et al.,* 1991). Yeasts are also important because some species can grow anaerobically, fermenting sugars to ethanol. When silages have substantial levels of sugars remaining after the lactic acid bacteria are inhibited by low pH, yeasts may develop and are the presumed cause of most high ethanol silages.

This ethanol production in silage not only decreases the amount of sugar available for lactic acid fermentation, but it can also have a negative effect on milk taste (Randby *et. al.,* 1999). Under aerobic conditions, many yeast species degrade the lactic acid to CO_2 and H_2O. The degradation of lactic acid causes a rise in silage pH, which in turn triggers the growth of many other spoilage organisms particularly once pH is above 4.5. (McDonald *et. al.*, 1991).

Yeast populations can reach up to 10^7 colony forming units per gram during the first weeks of ensiling; prolonged storage will lead to a gradual decrease in yeast numbers (Jonsson and Pahlow, 1984; Middelhoven and van Baalen, 1988; Driehuis and van Wikselaar, 1996). Factors that affect the survival of yeasts during storage are the degree of anaerobiosis and the concentrations of organic acids. The presence of oxygen enhances survival and growth of yeasts during storage (Jonsson and Pahlow, 1984), whereas high levels of formic or acetic acid reduce survival during storage (Driehuis and van Wikselaar, 1996; Oude Elferink *et al.*, 1999). Initial yeast activity appears to be enhanced in crops with a low initial pH (<5), for example, due to the addition of acid additives, and in crops with a high sugar content, such as potato, orange peel or sugar beet. These crops often result in silages high in ethanol and low in lactic acid (Ashbell *et al.*, 1987; Driehuis and van Wikselaar, 1996).

7. *Moulds*

Moulds are the filamentous fungi present on the crop. They are eukaryotic micro-organisms and are strictly aerobic. Mould-infested silage is usually easily identified by the large filamentous structures and coloured spores that many species produce. Moulds develop in parts of the silage where oxygen is present. During storage, this is usually only in the surface layers of the silage, but during aerobic spoilage (phase 4) the whole silage can become mouldy. They play an important role in the deterioration of silage when it is exposed to air. Mould species that regularly have been isolated from silage belong to the genera *Penicillium, Fusarium, Aspergillus, Mucor, Byssochlamys, Absidia, Arthrinium, Geotrichum, Monascus, Scopulariopsis* and *Trichoderma* (Nout *et al.*, 1993). Moulds not only cause a reduction in feed value and palatability of the silage, but can also have a negative effect on human and animal health. Mould spores are associated with lung damage and allergenic reactions (May, 1993). Other health problems are associated with mycotoxins that can be produced by moulds. Depending on the type and amounts of toxin present in the silage, health problems can range from minor digestive upsets, small fertility problems and reduced immune function, to serious liver or kidney damage, and abortions (Scudamore and Livesey, 1998). Some important mycotoxin-producing mould species are *Aspergillus fumigatus*, *Penicillium roqueforti*, and *Byssochlamys nivea*. *P. roqueforti*, a species which is acid tolerant and can grow at low levels of oxygen and high levels of CO_2, has been especially detected as the predominant species in different types of silages (Lacey, 1989; Nout *et al.*, 1993). Ensiling methods that minimize air ingress (e.g. good compaction and covering of the silo), and additives that prevent initiation of aerobic spoilage, will help to prevent or limit mould growth.

Summary of Microbiology and fermentation pathways of silage

Organisms	Conditions required	Major products/effects
Lactic acid bacteria	• Anaerobic condition • Wilting of crop is desirable • Chopping of crop for rapid establishment of lactic acid bacteria.	***Homofermentative pathway***: Glucose → 2 Lactic acid Fructose → 2 lactic acid Pentose → Lactic acid + Acetic acid ***Heterofermentative pathway***: Glucose → Lactic acid + Ethanol + CO_2 3 Fructose → Lactic acid +Methanol +Acetic acid + CO_2 Pentose → Lactic acid +Acetic acid

References

Ashbell, G., Pahlow, G., Dinter, B and Weinberg, Z.G. 1987. Dynamics of orange peel fermentation during ensilage. *J. Appl. Bacteriol.* 63:275-279.

Driehuis, F. and van Wikselaa, P.G. 1996. Effects of addition of formic, acetic or propionic acid to maize silage and low dry matter grass silage on the microbial flora and aerobic stability. In: D.I.H. Jones, R. Jones, R. Dewhurst, R. Merry, and P.M. Haigh (ed.) *Proc. 11th Int. Silage Conference*, Aberystwyth, UK. 8-11 September 1996. IGER, Aberystwyth, UK. p. 256-257.

Fenlon, D.R., Wilson, J and Weddell, J.R. 1989. The relationship between spoilage and *Listeria monocytogenes* contamination in bagged and wrapped big bale silage. *Grass Forage Sci.* 44: 97-101.

Gibson, J. 1965. Clostridia in silage. *J. Appl. Bacteriol.* 28:56-62.

Gibson, T., Stirling, A.C., Keddie, R.M and Rosenberger, R.F. 1958. Bacteriological changes in silage made at controlled temperatures. *J. Gen. Microbiol.* 19:112-129.

Giffel, M.C. te .1997. Isolation, identification and characterization of *Bacillus cereus* from the dairy environment. Ph.D. dissertation. Wageningen Agricultural University, The Netherlands.

Giffel, te M.C., Beumer, R.R., Slaghuis, B.A and Rombouts, F.R. 1995. Occurence and characterization of (psychrotrophic) *Bacillus cereus* on farms in the Netherlands. *Neth. Milk Dairy J.* 49:125-138.

Hammes, W.P., Weiss, N and Holzapfel, W. 1992. The Genera *Lactobacillus* and *Carnobacterium*. In: Balows, A., H.G. Trüper, M. Dworkin, W. Harder, K.H. Schleifer (ed.) *The Prokaryotes*. 2nd ed. Springer Verlag, New York, USA. pp 1535-1594.

Holzapfel, W.H., and Schillinger, U. 1992. The Genus *Leuconostoc*. p. 1508- 1534. In: Balows, A., H.G. Trüper, M. Dworkin, W. Harder, K.-H. Schleifer (ed.) *The Prokaryotes*. 2nd ed. Springer Verlag, New York, USA.

Jonsson, A. and Pahlow, G. 1984. Systematic classification and biochemical characterization of yeasts growing in grass silage inoculated with *Lactobacillus* cultures. *Anim. Res. Develop*. 20:7-22.

Kehler, W. and Scholz, H. 1996. Botulismus des Rindes. *Übersichten zur Tierernährung*, 24: 83-91.

Kleter, G., Lammers, W.L and Vos, A.E. 1982. The influence of pH and concentration of lactic acid and NaCl on the growth of of *Clostridium tyrobutyricum* in whey and cheese 1. Experiments in whey. *Neth. Milk Dairy J*., 36: 79-87.

Klijn, N., Nieuwenhof, F.F.J., Hoolwerf, J.D., van der Waals, C.B. and Weerkamp, A.H. 1995. Identification of *Clostridium tyrobutyricum* as the causative agent of late blowing in cheese by species-specific PCR amplification. *Appl. Environ. Microbiol*., 61:2919-2924.

Lacey, J. 1989. Pre- and post-harvest ecology of fungi causing spoilage of foods and other stored products. *J. Appl. Bacteriol*. 67(Suppl.): 11S-25S.

Lindgren S., Petterson, K., Kaspersson, A., Jonsson, A and Lingvall, P. 1985. Microbial dynamics during aerobic deterioration of silages. *J. Sci. Food Agr*. 36: 765-774.

May, J.J. 1993. Respiratory problems associated with work in silos. p. 283-290. In: Proc. NRAES National Silage Production Conference. Syracuse, USA. 23-28

McDonald P., Henderson, A.R. and Heron, S.J.E. 1991. The Biochemistry of Silage. 2nd edition. Chalcombe Publications, Marlow, Bucks, UK.

McPherson, H.T., and Violante, P. 1966. Ornithine, putrescine and cadaverine in farm silages. *J. Sci. Food Agr*. 17:124-127.

Middelhoven, W.J. and van Baalen, A.H.M.1988. Development of the yeast flora of whole-crop maize during ensiling and subsequent aerobiosis. *J. Sci. Food Agr*. 42:199-207.

Moran, J.P., Pullar, D. and Owen, T.R. 1993. The development of a novel bacterial inoculant to reduce mould spoilage and improve the silage fermentation in big bale silage. In: P. O'Kiely, M. O'Connell, and J. Murphy (ed.) Silage Research 1993, Proc. 10th Int. Conf. Silage Res., Dublin, Ireland. 6-8 Sept 1993. Dublin City University, Dublin, Ireland. p. 85-86.

Muck, R.E and Pitt, R.E. 1994. Aerobic deterioration in corn silage relative to the silo face. Transactions of the ASAE. 37(3): 735-743.

Nout, M.J.R., Bouwmeester, H.M., Haaksma, J. and van Dijk, H. 1993. Fungal growth in silages of sugarbeet press pulp and maize. *J. Agr. Sci*. 121: 323-326.

O'Kiely, P., Turley, T. and Rogers, P.A.M. 1999. Exposure of calves to nitrogen dioxide in silage gas. *Vet. Rec*. 144: 352-353.

Oude Elferink, S.J.W.H., Driehuis, F., Krooneman, J., Gottschal, J.C and Spoelstra, S.F. 1999. *Lactobacillus buchneri* can improve the aerobic stability of silage via a novel fermentation pathway, the anaerobic degradation of lactic acid to acetic acid and 1,2-propanediol.. In: T. Pauly (ed.) Proc. 12th Int. Silage Conference, Uppsala, Sweden, 5-7 July. 1999. Swedish University of Agricultural Sciences, Uppsala, Sweden. p. 266-267

Pahlow, G., Muck, R.E and Driehuis, F. 2003. Microbiology of ensiling. In: Buxton, D.R.; Muck, R.E.; Harrison, J.H. (Eds.) *Silage Science and Technology*. Madison: American Society of Agronomy, Crop Science Society of America, Soil Science Society of America. p. 31-93.

Phillip, L.E and Fellner, V. 1992. Effects of bacterial inoculation of high moisture ear corn on its aerobic stability, digestion, and utilization for growth by beef steers. *J. Anim. Sci.* 70: 3178-3187.

Randby, Å.T., Selmer-Olsen, I and Baevre, L. 1999. Effect of ethanol in feed on milk flavor and chemical composition. *J. Dairy. Sci.* 82: 420-428.

Sanaa, M., Poutrel, B., Menard, J.L. and Serieys, F. 1993. Risk factors associated with contamination of raw milk by *Listeria monocytogenes* in dairy farms. *J. Dairy Sci.* 76: 2891-2898.

Schlegel, H.G. 1987. General Microbiology. 6th ed. Cambridge University Press, Cambridge, UK.

Schleifer , K.H and Ludwig, W. 1995. Phyogenetic relationships of lactic acid bacteria. In: Wood, B.J.B. and Holzapfel, W.H. (ed.) *The Genera of Lactic Acid Bacteria.* Blackie Academic & Professional, London, UK. 7-18.

Scudamore, K.A., and Livesey, C.T. 1998. Occurence and significance of mycotoxins in forage crops and silage, a review. *J. Sci. Food Agr.* pp 77: 1-7.

Schlegel, H.G. 1987. *General Microbiology.* 6th ed. Cambridge University Press, Cambridge, UK.

Spoelstra, S.F. 1983. Inhibition of clostridial growth by nitrate during the early phase of silage fermentation. *J. Sci. Food Agr.* 34: 145-152.

Spoelstra, S.F. 1985. Nitrate in silage. A review. *Grass Forage Sci.* 40: 1-11.

Spoelstra, S.F. 1987. Degradation of nitrate by enterobacteria during silage fermentation of grass. *Neth. J. Agr. Sci.* 35: 43-54.

Spoelstra, S.F., Courtin, M.G. and van Beers, J.A.C. 1988. Acetic acid bacteria can initiate aerobic deterioration of whole crop maize silage. *J. Agr. Sci.Camb.* 111: 127-132.

Teuber, M., Geis, A and Neve, H. 1992. The Genus *Lactococcus*. In: Balows, A., H.G. Trüper, M. Dworkin, W. Harder, K.-H. Schleifer (ed.) *The Prokaryotes*. 2nd ed. Springer Verlag, New York, USA pp 1482-1501.

Van, O, M. and Dulphy, J.P. 1996. Voluntary intake and intake control of grass silage by ruminants. *Reprod. Nutr. Develop.* 36: 113-135.

Van, O, M, A.M.van Vuuren, and S.F. Spoelstra 1997. Mechanisms of adaptation in sheep to overcome silage intake depression induced by biogenic amines. *Brit. J. Nutr.* 77: 399-415.

Weiss, N. 1992. The Genera *Pediococcus* and *Aerococcus*. In: Balows, A., H.G. Trüper, M. Dworkin, W. Harder, K.-H. Schleifer (ed.). *The Prokaryotes*. 2nd ed. Springer Verlag, New York, USA. pp 1502-1507.

Wiedmann, M., Czajka, J., Bsat, N., Bodis, M., Smith, M.C., Divers, T.J. and Batt, C.A. 1994. Diagnosis and epidemiological association of *Listeria monocytogenes* strains in two outbreaks of listerial encephalitis in small ruminants. *J. Clin. Microbiol.* 32: 991-996.

Wieringa, G.W. 1958. The effect of wilting on butyric acid fermentation in silage. *Neth. J. Agr. Sci.* 6: 204-210.

Woods, L.F.J., Wood, J.M. and Gibbs, P.A. 1981. The involvement of nitric oxide in the inhibition of the phosphoriclastic system in *Clostridium sporogenes* by sodium nitrite. *J. Gen. Microbiol.* 125: 339-406.

Woolford, M.K. 1984. The Silage Fermentation. Microbiological Series, 14, Marcel Dekker, Inc., New York.

CHAPTER 6

Chemistry of Silage

Silage fermentation is a complex process involving biochemical changes by microbial populations and the ensiling environment. The chemistry and microbiology of silage fermentation are linked, but the microorganisms are the dominant factor in the fermentation process, and they have the main influence on the quality of the end product. Many species of microorganisms are involved in the fermentation process, and their relative importance varies according to the prevailing conditions. As these conditions change during the fermentation process, the microorganism population changes accordingly. Lactic acid bacteria and clostridia are two main categories of microbes responsible for biochemical changes in silage.

The main chemical components that undergo changes during silage fermentation are water soluble sugars, organic acids, and nitrogen compounds.

A. Carbohydrates

The water soluble carbohydrates are important sources of energy for those organisms which are responsible for silage fermentation. Fructose, glucose, sucrose and fructosans are the principal water soluble sugars in forage crops, which undergo fermentation during ensiling. Sucrose and fructosans are rapidly hydrolyzed to their monomers at the time of harvesting. Fructans have a minor role in silage fermentation but are the major storage carbohydrates in grasses in temperate region. Grasses of tropical and subtropical origin accumulate starches in their vegetative tissues and these are stored primarily in the leaves. The relative amounts of the water soluble carbohydrates in grasses are influenced by the leaf to stem ratio, time of day, light intensity and temperature (Smith, 1973). The water soluble carbohydrate concentration of grasses is very variable, ranging from 2g/kg DM in some tropical species to over 300g/kg DM in some cutivars of rye grass. In grasses fructose and glucose accounts for 10 to 30 g/kg and sucrose from 20 to 80 g/kg DM (McDonald *et.al.*, 1961). While fructose and glucose are the principal fermentable substrates in a grass crop, fructose is the most predominant carbohydrate in the grass. Ratio of glucose to fructose ranges from 1.1 to 3.9. This ratio is of importance in silage when heterofermentative lactic bacteria are predominant, since they ferment the fructose to acid less efficiently than do their homofermentative counterparts and less efficiently than glucose (McAllan and Phipps, 1977). Both sucrose and fructosans are rapidly hydrolysed to their component monomers at the time of harvest. Pentoses are found to a limited extent in forage crops, appearing as a result of hemicelluloses hydrolysis. Pentoses are femented by both homo and heterofermentative pathway to yield each of lactic acid and acetic acid. Structural carbohydrates (e.g. cellulose) are of little significance to the silage fermentation although hemicelluloses may make a minor contribution to the provision of water soluble carbohydrates (pentoses) through the action of hemicellulases (Dewar *et al.*, 1963). It is unfortunate that pentoses are not more abundant but they are fermented to lactic acid by both homefermatative and heterolactic fermentative lactic acid bacteria.

The water soluble sugars are converted mainly to mixtures of organic acids, whose composition depends on the types of microorganisms involved and on the prevailing conditions. Lactic acid bacteria are the most important microorganisms for the ensiling process. The lactic acid bacteria which are facultative anaerobes are normally present on growing crops in small numbers but usually multiply rapidly after harvesting particularly if the crop is chopped or lacerated. The lactic acid bacteria comprise a large group of bacteria, encompassing species belonging to six genera. They primarily ferment sugars to lactic acid, but they also produce some acetic acid, ethanol, carbon dioxide and other minor products. The pathway through which substrates are fermented by lactic acid bacteria differ according to species of microorganisms and may alter according to the nature of the substrate. Depending upon their fermentative ability the lactic acid producing bacteria can be divided into two categories i.e. Homofermentative bacteria (*Lactobacillus plantarum, Pedicoccus pentosaceus* and *Enterocccus faecalis*) and Heterofermentative bacteria (*Lactobacillus breveris* and *Leuconostoc mesenteroides*).

I. Homolactic fermentation

The homofermatative pathway follows the Embden-Mayerhoff-Paranass (EMP) scheme in the steps to pyruvate followed by dehydrogenation of this product to yield lactate. One molecule of glucose or fructose is fermented to two moles of lactate (Figure.1). Under anaerobic conditions the homofermentative lactic acid bacteria produce only lactic acid from hexoses, such as glucose and other six-carbon sugars. The lactic acid lowers the pH more quickly, so reducing the activity of proteolytic enzymes, and conserving protein (McDonald, 1981). The low pH also stops the growth of other undesirable anaerobic bacteria, such as *enterobacteria, clostridia* and *listeria*. Homofermentative fermentation, which is the more desirable, yields lactic acid, which is a stronger acid than acetic acid, and does not produce carbon dioxide, which represents a loss of DM. Thus, homofermentative fermentation results in little or no loss of DM and only a small energy loss. The homomfermentative lactic acid bacteria are more efficient at producing lactic acid from hexose sugars than are the heterofermentative bacteria.

II. Heterolactic fermentation

In case of heterofermentative fermentation, hexoses are fermented by heterofermentative lactic acid bacteria to lactic acid, ethanol, acetic acid, and carbon dioxide. The primary route through which hexoses are fermented by the heterolactic mechanism is the hexose monophosphate pathway. If glucose is the substrate, one mole of the glucose is asymmetrically split to yield one mole each of lactate, ethanol, and carbon dioxide. On the other hand, if fructose is used, then every three moles of fructose will yield one mole each of lactate, acetate, and carbon dioxide and two moles of mannitol. Fructose is reduced to lactic acid, mannitol, acetic acid and carbon dioxide. Therefore, the heterolactic fermentation will be even less efficient in these terms if fructose is more abundant than glucose.

In case of pentoses (arabinose or xylose) both homo and heterofermentative lactic acid bacteria share a common pathway for the fermentation to yield each of lactic acid and acetic acid. Although

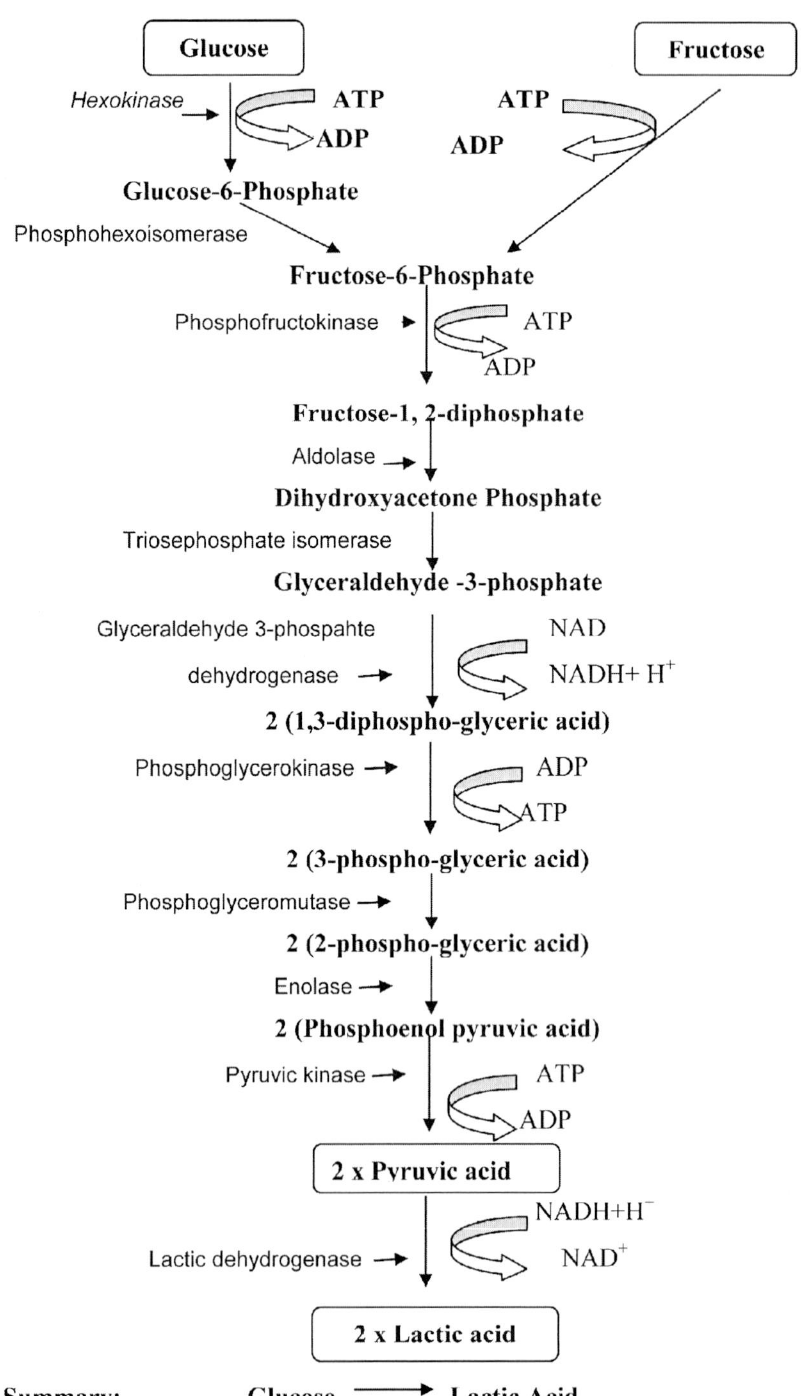

Fig. 1: The homolactic fermentation of glucose and fructose

in terms of acidification this fermentation is not as effective as the homolactic fermentation of hexoses, because acetic acid being relatively weaker than lactate. The heterolactic fermentation is less efficient in terms of acid production than its homolactic counterpart. However, even a totally heterofermentative process will rarely result in more than a 5% loss of DM, and the gross energy loss is negligible. Summary of heterolactic fermentation are as follows:

Glucose → Lactic acid + Ethanol + Carbon dioxide

Pentose → Lactic acid + Acetic acid

Fructose + Water → Lactic acid + Acetic acid+ Mannitol+ Carbon dioxide

Clostridial spoilage

Clostridia are a group of undesirable bacteria that grow in silage, especially in moist silage (below 30% DM) under anaerobic conditions. They are spore-forming bacteria, which enables them to survive for a long time, even under bad conditions. Soil and manure are the most common sources of clostridia; therefore great care should be taken to avoid contamination of the forage with soil during harvesting.

Clostridia can be divided into two groups. The first group comprises saccharolytic species such as *Clostridium butyricum*, which ferment sugars and lactic acid to produce butyric acid, gaseous carbon dioxide and hydrogen, via the following pathways (Klijn *et al.*, 1995):

Sugar →Butyric acid + Carbon dioxide + Hydrogen

Lactic acid→ Butyric acid + Carbon dioxide + Hydrogen

Production of carbon dioxide and hydrogen gas represent losses of digestible DM and energy. Also, butyric acid is a weaker acid than lactic acid, and clostridia produce only one molecule of butyric acid from two molecules of lactic acid or sugar. Such fermentation reduces the DM content by about 50%, and the energy content by almost 20%; it is referred to as secondary fermentation, and is accompanied by increasing pH. The weakness of the acid and the production of ammonia both raise the pH, and high levels of ammonia also depress the animals' intake.

Clostridial fermentation is characterized by a foul smell, indicative of the production of components such as putrescine and cadaverine; its many other negative effects include poor preservation of forage and silage, high pH, high DM losses, high ammonia-N production, low intake, and butyric acid production. Thus, it is important to prevent clostridial bacterial activity, which is encouraged mainly by low DM content and low pH, and can therefore be limited by ensiling forage with DM of at least 30%, and by rapidly achieving a low pH in the silage. All factors that contribute to these two aims, such as type of crop, wilting rate, buffer capacity, sugar content, and soil contamination, will also influence the activity of clostridia.

B. Nitrogenous substances

Proteins are the main nitrogenous compounds in herbage. In fresh forage normally major portion of nitrogen exists in the form of true protein, which in some cases may vary between 80-90% and the remaining 10-20% is in the form of non-protein nitrogen which includes free amino acids, amines, amines, nucleotides, chlorophyll, nitrates and ammonia. Arginine and glutamic acid are predominant amino acids of forage crop proteins together with significant amount of lysine, aspartic acid, alanine and glycine. Amino acid composition of herbage protein is influenced by plant species, stage of maturity and the quantity of nitrogenous fertilisers applied. Grass proteins are particularly rich in arginine. Nitrates may also be present in the forage depending on the amount of nitrogenous fertilizer applied. The quantity of ammonia nitrogen is usually very low in fresh herbage, often less than 1.5% of the total nitrogen (Jones and Wilson, 1987; McDonald, 1991).

In the case of the nitrogen compounds, rapid and extensive proteolysis occurs directly after cutting, mainly through the action of plant enzymes. This process continues during the ensiling, and its duration depends strongly on the ensilability of the forage, and conditions in the silage. It can be expected that over 50% of the true protein present in the forage will have disappeared from the silage when the fermentation process is over (Bergen *et al.*, 1974). Moisture content of the forage is also an important factor in proteolysis during wilting. In the presence of sufficient moisture, proteolysis may be very rapid but its content is influenced by the time taken to remove moisture. Proteolysis may be encouraged by wilting the crop. Therefore, rapidly dried crops have little protein breakdown. Proteolysis continues during ensilage. The extent of proteolysis depends largely on the rapidity with which acid conditions are established. Thus, the rate of acidification occurring in the silage is important for the conservation of protein. Therefore, acidification by inorganic acids (hydrochloric acid and sulphuric acid) is prevents proteolysis during ensiling (Virtanen, 1933).

Amino acids usually do not undergo change during ensilage and good ensiling conditions. Lactic acid bacteria have the activity of protease, deaminase and decarboxylase, but their role is not very significant in the total degradation of protein and catabolism of amino acid. Proteolytic clostridia (*Clostridium perfingens, Clostridia bifermentans*) convert amino acids through deamination, decaboxylation and coupled oxidation/reduction reaction (stickland) to a variety of products including acetic acid, ammonia, amines and volatile fatty acids. These are as follows:

Deamination

Lysine →Acetic acid + Butyric acid + 2NH_3

Glutamic acid → Acetic acid + Pyruvic acid + NH_3

Decarboxylation

Arginine →Putriscine + CO_2

Glutamic acid → gama aminobutyric acid+ CO_2

Lysine →Cadaverine + CO_2

Histidine →Histamine + CO_2

Oxidation/reduction (stickland)

Alanine + 2 Glycine → 3 Acetic acid + 3NH_3 + CO_2

Production of carbon dioxide and hydrogen gas represent losses of digestible DM and energy, and production of ammonia is a loss of protein. Clostridial fermentation is characterized by a foul smell, indicative of the production of components such as putrescine and cadaverine; its many other negative effects include poor preservation of forage and silage, high pH, high DM losses, high ammonia-N production, low intake, and butyric acid production. Thus, it is important to prevent clostridial bacterial activity, which is encouraged mainly by low DM content and low pH, and can therefore be limited by ensiling forage with DM of at least 30%, and by rapidly achieving a low pH in the silage. All factors that contribute to these two aims, such as type of crop, wilting rate, buffer capacity, sugar content, and soil contamination, will also influence the activity of clostridia. Clostridial silages become unpalatable primarily from nitrogenous proteolysis in the production of unpalatable amines, amides, and other nitrogenous end products.

C. Organic acid

The predominant organic acids in forage are citric acid, malic acids, aconitic with lesser amounts of quinic, succinic and oxalic acid. They constitute 2-6% of the DM in grasses and 6-8% of that in legumes. Organic acids are present in maximum concentrations in immature forages and decline with age and increasing maturity of forages. They are apt to be present in considerable amounts when forage is harvested for silage. Most of theses acids are di- or tricaroxylic exist as mono-or dibasic salts of potassium and calcium in the plant. The salts of these acids have a buffering action during the fermentation process, but the plant enzymes quickly metabolize them during the ensilage and wilting. The disappearance of organic acids during ensilage is due to their fermentation by lactic acid bacteria. It was observed that homofermentative lactic acid bacteria could ferment citrate to lactate, formate, acetate, acetoin, 2,3-butane diol and carobon dioxide. Acetoin in silage may also be derived from serine by deamination. At a low pH lactate is the predominant product and as the pH increases, larger proportions of formate and acetate appear until at pH 7 and above.

D. Pigments

The colour of silage changes from green to light brown. This change is the result of the action of fermentation acids on chlorophyll which yields a magnesium free pigment known as "phacophytin". The loss of carotene is inevitable. Carotene, a pro-vitamin is destroyed to some extent during ensilage. In good quality silage, loss of this pigment may be less than and reduced further by restricting fermentation 30%. However, in poor quality silage the loss of carotene may be 100%. Carotene loss is encouraged by air ingress into the silo, therefore, losses of carotene are minimised by restricting fermentation and imposing anaerobic conditions.

Summary of important pathways during ensiling process

Pathway		Substrate recovery (%)	
Substrate	*End product*	*Dry matter*	*Energy*
	Homofermentative		
1 glucose	2 Lactic acid	100	97
	Heterofermentative		
Glucose	Lactic acid + Ethanol + CO_2	76	97
3 Fructose	Lactic acid + 2 Mannitol + Acetic acid + CO_2	95	98
Clostridial			
2 Lactic acid	Butyric acid + 2 CO_2 + 2 H_2	49	81
2 Alanine	Propionic acid + Acetic acid + 2 NH_3 + CO_2	78	81
Yeast			
Glucose	2 Ethanol + 2 CO_2	51	97

References

Bergen, W.G., Cash, E.H. and Hendersin, H.E. 1974. Changes in nitrogenous compounds of whole corn plant during ensiling and subsequent effects of dry intake by sheep. *Journal of Animal Science.* 39:629-637.

Dewar, W.A., McDonland, P. and Wittenburry, R. 1963. The hydrolysis of grass hemicellulose during ensiling. *Journal of the Science of Food and Agriculture.*14:411-417.

Gregory, P.H., Lacey, M.E., Festentein, G.N. and Skinner, F.A. 1963. Microbial and biochemical changes during the molding of hay. *J. Gen. Microbiology.* 33:147 174.

Jones, D I H and Wilson A.D. 1987. Nutritive quality of forages. In: *The Nutrition of Herbivores.* Hacker JB and Ternouth JH (eds). Academic Press, Sydney.

Klijn, N., Nieuwenhof, F.F.J., Hoolwerf, J.D., van der Waals, C.B and Weerkamp, A.H. 1995. Identification of *Clostridium tyrobutyricum* as the causative agent of late blowing in cheese by species-specific PCR amplification. *Appl. Environ. Microbiol.* 61:2919-2924.

McAllan, A.B and Phipps, R.H. 1977. The effect of sample date and plant density on the carbohydrate content of forage maize and changes that occur in silage. *Journal of Agricultural Sciences.* 89: 589-597.

McDonald, P. 1981. *The Biochemistry of Silage.* John Wiley and Sons, New York.

McDonald P., Sterling, A.C., Henderson, A.R., Dewar, W.A., Stark, G.H., Davie, W.G., Macpherson, H.T., Reid, A.M and Slater, J. 1960. Studies of Ensilage. *Edinberg School of Agriculture,* Technical Bulletin No. 24. pp. 34.

McDonald P., Henderson, A.R and Heron, S.J.E. 1991. The Biochemistry of Silage. 2nd edition. Chalcombe Publications, Marlow, Bucks, UK.

Smith, D. 1973. The non-structural carbohydrates. In: *Chemistry and Biochemistry of Herbage.* G.W. Butler and R.W.Bailey (Eds.), Academic Press, London pp 639.

Virtanen, A I V. 1933. The AIV method of processing fresh fodder. *Empire Journal of Experimental Agriculture.* 1:143-145.

CHAPTER 7

Silage Additives

Silage fermentation is a dynamic process that is affected by variety of factors. The primary goal of making silage is to maximize the preservation of original nutrients in the forage crop for feeding at a later date. Unfortunately, fermentation in the silo is a much uncontrolled process usually leading to less than optimal preservation of nutrients. In order to assist in the fermentation process, various silage additives have been used to improve the nutrient and energy recovery in silage, often with subsequent improvements in animal performance (Henderson, 1993; Bolsen *et al.*, 1996; Muck and Kung, 1997; Kung and Muck, 1997; Muhlbach, 2000). Their main functions are to either increase nutritional value of silage or improve fermentation so that storage losses are reduced. The benefits obtained from silage additives depend upon their influence on the silage fermentation process. These benefits are usually measured by the reduction in fermentation losses and/or improvement in silage quality and feeding value.

Silage additives or preservatives function in the following ways:

- Alter the rate, amount and kind of acid production
- Acidify the silage
- Inhibit bacterial and mold growth
- Culture silage (inoculants) to stimulate acid production
- Increase nutrient content of the silage

Silage additives have been classified into various categories that generally include

1. Stimulants of fermentation (microbial inoculants, enzymes, fermentable substrates),
2. Inhibitors of fermentation (acids, other preservatives), and
3. Nutrient additives (ammonia, urea, minerals).

A. Fermentation stimulants

This group of additives operates by encouraging the fermentation. Assuming good harvesting and ensiling techniques, initial silage fermentation can still be sub-optimal due to a lack of sufficient numbers of suitable lactic acid bacteria (LAB) or a lack of sufficient amounts of suitable water soluble carbohydrates (WSC), or both. The amount of WSC necessary to obtain sufficient fermentation depends on the DM content and the buffer capacity of the crop. Weissbach and Honig (1996) characterized the relation between these factors as follows,

$$FC = DM\,(\%) + 8\,WSC/BC$$

Where:

FC = fermentation coefficient

DM = dry matter content

WSC = water-soluble carbohydrates

BC = buffer capacity.

Forages with insufficient fermentable substrate or too low DM content have FC <35. In these forages, sufficient fermentation can only be achieved if the sugar content of the material is increased, either by adding sugars directly (e.g. molasses) or by adding enzymes that release extra sugars from the crop. In forages with an FC of 35 or more, sufficient fermentable substrate is available. Also, adding suitable LAB can accelerate and improve the ensiling process. In high DM silages with reduced water availability, the presence of suitable, osmo-tolerant LAB could become a limiting factor in the ensiling process. It has been shown that these bacteria represent only a small percentage of the indigenous microflora on forage crops (Pahlow and Weissbach, 1996). Forages with DM content above 50% are considered difficult to ensile (Staudacher *et al.*, 1999). The formula of Weissbach and Honig (1996) is not applicable for crops with low nitrate content, such as extensively managed grasses and immature whole-crop cereals, because these crops are more liable to clostridial fermentations than crops with moderate nitrate content (Spoelstra, 1985,1987). Inoculants that increase lactic acid fermentation might be useful to inhibit clostridial activity. The minimum number of LAB required to inhibit clostridial activity was found to be at least 100 000 colony-forming units per gram of fresh crop (Weissbach and Honig, 1996; Kaiser and Weiss, 1997).

Stimulation of fermentation can be achieved by the following three ways:

(a) Addition of efficient lactic acid producing micro-organism

(b) Addition of a substrate which will promote a lactic acid fermentation or

(c) Enzymatiic breakdown of forage plant components rich in fermentable substrate

A.1. Fermentable substrates

The incorporation of easily fermentable feed ingredients such as sugar or molasses to low-DM, sugar-limited tropical forages is a way to improve silage fermentation. Feed-grade products such as grains in general and processed by-products such as maize or sorghum meal, rice bran, cassava meal, citrus pulp, etc., can also be used as additives, partly to provide fermentable substrate, but also to direct the course of fermentation by absorbing excessive moisture.

Molasses which is a by-product of the sugar cane and sugar beet industries can be used as a source of water soluble carbohydrates for lactic acid bacteria thus causing the desired reduction in pH. The by-product has 75 to 80 percent moisture, and about 70% water soluble carbohydrate content of dry matte of which sucrose is the main component. Molasses provides a relatively cheap source of fermentable carbohydrate for lactic acid bacteria. Cane molasses (75% DM) has been widely used, added up to 10% w/w to provide fast fermentable carbohydrate for the ensilage of tropical herbages. Due to its viscosity, it is difficult to apply and should be diluted, preferably with a small volume of warm water to minimize seepage losses. When applied to tropical grasses, molasses should be used in relatively high concentrations (4 to 5%). With crops of very low DM content, a considerable proportion of the additive may be lost in the effluent during the first days of ensilage (Henderson, 1993). Sugar cane molasses added at the rate of 3% (w/w, fresh basis) to Napier grass (12.9% DM, 6.6% WSC) produced silages of reasonably good fermentation quality, reducing, however, the nutrient recovery from the silo, as compared to formic-acid-treated silage (Boin, 1975). Molasses is an effective silage additive in terms of promoting lactic fermentation, reducing silage pH, discouraging a clostridial fermentation and proteolysis, and generally decreasing organic matter losses. It is of particular benefit when applied to forage crops low in fermentable carbohydrates for lactobacilli. Keady (1996) reviewed the published literature on molasses as silage additives and concluded that molasses treatment improved silage preservation, but did not significantly alter the silage digestibility or animal performance although silage DM intake was improved. The additive has been shown to increase the dry matter and lactic acid, and to reduce the pH and ammonia levels in treated silage. Adding molasses at 200-400 lbs./ wet ton of crop silage (approximately 45% dry matter) usually will increase fermentation noticeably and result in good quality silage. Adding 50-100 lbs. molasses/wet ton will increase fermentation and usually increase bunk life of silage. Molasses should not be added to wet silage (<35% dry matter) because of increased seepage loses. Elephant grass was treated with 4% molasses and ensiled in 4-kg polythene bags, with the resulting silage having lower pH and ammonia-N than the control silage (Tosi *et al.*, 1995). There is no benefit of adding molasses to maize silage.

A.2. Microbial inoculants

Inoculants are the most common silage additives. Since silage is a product resulting from the action of bacterial enzymes on the material stored, attempts have been made to alter or regulate silage fermentation through the addition of bacteria, yeasts and molds. It is known that growing crops are often poor source of lactic acid bacteria and that some strains of theses organisms are not ideally suited for ensiling purposes. Microbial inoculants contain one or more of beneficial bacteria which have their ability to dominate the fermentation. Lactic acid is the most preferred end product of silage fermentation. Therefore, adding suitable lactic acid bacteria can accelerate and improve the ensiling process. The primary purpose for adding bacterial inoculants is to increase the number of lactic acid-producing bacteria, thus encouraging more lactic acid production and well-preserved forage mass. Lactic acid bacteria, propionic acid bacteria and yeast are major microbial inoculants used as fermentation stimulant.

Homolactic acid fermentation that produces only lactic acid would be the desirable fermentation because of the high energy and dry matter recoveries. Thus, the concept of adding a microbial inoculant to silage was to add fast growing homofermentative lactic acid bacteria in order to dominate the fermentation resulting in higher quality silage. Some of the more common homolactic acid bacteria used in silage inoculants include *Lactobacillus plantarum, Lactobacillus acidophilus, Pediococcus acidilactici, Pediococcus pentacaceus, and Enterococcus faecium.* Some *Pediococcus* strains are more tolerant of high dry matter conditions than are *Lactobacillus* and have a wider range of optimal temperature and pH for growth. Silage inoculant also inhibits aerobic spoilage. Heterolactic lactobacilli like *Lactobacillus buchneri* is a very effective inhibitor of aerobic spoilage. The inhibition of spoilage is mainly due to the capability of *Lactobacillus buchneri* to anaerobically degrade lactic acid to acetic acid and 1, 2-propanediol, which in turn causes a significant reduction in yeast in silage (Driehuis *et al.*, 1996 ; Ranjit *et al.*, 1998; Oude-Elferink *et al.*, 1999). Recently developed inoculants use *Lactobacillus buchneri* 40788, a group of bacteria which restrict the growth of spoilage type organisms such as yeasts and moulds. Their mode of action is thought to be due to secondary metabolites called "buchnericides". Of the various types of fermentative bacteria available, homofermentative lactic acid bacteria are the first choice; and the vast majority of commercially available inoculants contain these bacteria.

Before a culture can be considered suitable for inclusion in an inoculant, it must satisfy a number of criteria (Whittenbury, 1961) as follows:

i. It must have a high growth rate and be able to compete with and dominate other organisms likely to occur in silage.

ii. It must be homofermentative.

iii. It must be acid tolerant and produce a final pH of 4.0 quickly.

iv. It must be able to ferment glucose, fructose, sucrose, and preferable fructosans and pentosans.

v. It must not produce dextran from sucrose or mannitol from fructose.

vi. It should have no action on organic acids (such as citric, malic and succinic would enhance buffering and thereby, hinder acidification).

vii. It should have a good growth temperature range, preferably up to 50°C, in order to survive any rise in temperature during the early stages of ensilage.

Efficacy of microbial inoculant depends upon the inoculation rate as well as the presence of adequate concentrations of sugar. The organism(s) from microbial inoculants must be present in sufficient numbers to effectively dominate the fermentation. The minimum number of lactic acid bacteria should be at least 10^5 colony-forming units per gram of fresh crop (Kaiser and Weiss, 1997). Consequently, a number of commercial inoculants containing freeze-dried cultures of homofermentative lactic acid bacteria have become available for application to harvested forage. Most microbial inoculants are available in powder or granular form. Inoculants applied in the dry form are often mixed with calcium carbonate (limestone), dried skim milk, sucrose or other carriers. These products can be applied by hand or by solid metering devices as per manufacturer's recommendations. Inoculants to be applied in the liquid form come as dried powders and are mixed with water just prior to use. (Use of chlorinated water may be detrimental to the inoculant if levels exceed more than 1.5 to 2 ppm.) Application can be with a simple watering can by weighing the incoming forage load and adjusting application based on the average unloading time. A better method is to use a metered liquid sprayer to evenly disperse the inoculants on the forage. Unused liquids should be discarded after a period of 24 to 48 h because bacterial numbers begin to decline. Inoculants applied in a liquid form may be more advantageous because the bacteria are added with their own moisture to help speed up fermentation. Storage is an important aspect of a high quality inoculant that contains live microorganisms. Some inoculants require refrigeration or freezing for optimum storage. Moisture, oxygen and sunlight can decrease the stability of inoculants resulting in lower viable counts and a product that does not meet label guarantees.

Propionibacteria are also used as silage inoculant which are able to convert lactic acid and glucose to acetic and propionic acids that are more antifungal than lactic acid. Addition of *propionibacteria* improves aerobic stability and prevents the growth of molds and yeast in high moisture silage (Florez-Galaraza *et al.*, 1985; Dawson, 1994; Bolsen *et al.*, 1996).

A.3. Enzymes

Enzymes are proteins that assist in metabolic processes. A variety of enzymes, particularly those that digest plant fiber and starch have been used as silages additives. Forage plants have a large reserve of carbohydrate as polysaccharides. Enzymes are added to silage to degrade polysaccharides in cell wall of plants to fermentable water-soluble carbohydrates for use by lactic acid bacteria because these organisms cannot use polysaccharides as an energy source to make lactic acid. Cellulases and hemicellulases are the most widely used enzyme additives. The enzymes appear to be most efficient when added to young herbage ensiled at low dry matter content.

There are two primary reasons for adding fiber-digesting enzymes to silage. First these enzymes could partially digest the plant cell walls (cellulose and hemicellulose) yielding soluble sugars which could be fermented by lactic acid bacteria to lower the silage pH. This would stimulate silage fermentation and improve fermentation quality by increasing the rate and extent of decline in pH, increasing the concentration of lactic acid, improving the lactic acid: acetic acid ratio (which is indicative of greater efficiency of fermentation), and hence reduce DM losses. A faster decline in pH would also limit degradation and deamination of forage proteins and reduce ammonia production. Second, partial digestion of the plant cell wall may improve the rate and/or extent of digestibility. In order for the first event to take place the rate of cellulose hydrolysis must coincide with early growth of lactic acid bacteria.

B. Nutrient additives

Certain crops are deficient in dietary components essential for ruminants. The nutritional quality of these crops can be improved by supplementation with specific additives at the time of ensiling. Various materials added to silage will increase nutritive value to the extent that they themselves contain nutrients. Materials such as cereal grains, molasses, dry forages, limestone, urea and anhydrous ammonia are examples of nutrient additions to silage.

B.1. Nitrogen additives

Urea and anhydrous ammonia, non-protein nitrogen (NPN) sources, are sometimes used as nitrogen additives for silage. Anhydrous ammonia is basic in nature and immediately after application will elevate the pH of silage. Afterwards, the pH slowly declines via normal fermentation, but fermentation will not be as extensive as untreated corn silage

Ammonia additions can result in

a) Addition of an economical source of crude protein (Huber *et al.,* 1979; McDonald *et al.,* 1991).

b) Aerobic stability thus, prolonged bunk life during feeding (Britt and Huber, 1975);

c) Less molding and heating during ensiling; and

d) Decreased protein degradation in the silo (Johnson *et al.,* 1982).

The recommended application rate for anhydrous ammonia is 7-9 lbs./ton of wet silage. Anhydrous ammonia should be added using a cold-flow system to reduce losses during application. If application rate exceed the maximum recommended rates because both urea and ammonia buffer silage and will raise the final pH. Silage with high final pH is not well-preserved. Recently, research has shown that adding 2-4 lbs anhydrous ammonia/wet ton of hay crop silage reduces fermentation losses. The amount of ammonia added to hay crop silage must not exceed 4 lbs/ton or fermentation will be inhibited severely. Ammoniating maize and hay crop silage at the recommended rates should reduce fermentation losses, increase bunk life, and may enhance consumption of the silage by animals. A benefit of adding NPN to maize silage includes increasing the CP content by 2-6 percentage units. Ammoniated or urea-treated maize silage usually has CP concentrations of 12-13% (DM basis). This can result in substantial savings by reducing the need for protein supplementation. Another benefit of adding NPN to silage crops is that it reduces plant protein destruction during fermentation. During normal fermentation, much of the plant protein is broken down into amino acids or small peptides. These compounds are degraded rapidly in the rumen when cattle or sheep are fed the silage. Saving plant proteins from breakdown during ensiling increases the amount of undegradable or bypass protein in the silage. This could be very advantageous with high quality grass or alfalfa silage. Anhydrous ammonia should not be added to maize forage if the DM content is above 40-42% because fermentation is restricted in drier material and binding of ammonia will be less; thus normal fermentation may be disrupted. Where forage DM is above 40-42%, water-ammonia mixes or molasses-ammonia mixes should be used. Ammonia is a hazardous gas and should be handled with care. Eye protection should be worn when making connections to pressurized tanks. Water should be available at all times. Ammonia is also corrosive to zinc, copper and brass. Therefore storage of ammonia-treated forage in zinc coated steel silos is not recommended.

Urea has been added to maize silage as an economical source of crude protein. Urea has 280% CP and anhydrous ammonia contains 512% CP. Therefore, relatively small additions of these compounds can markedly increase the CP content of silage. Maize silage is extremely low in CP (average 8%, DM basis); whereas hay crop silage may contain more than 20% CP. The recommended application rate of urea for maize silage of normal DM content (approximately 35%) is 0.5% of fresh forage. Whenever ammonia or urea is added to the diet, special attention should be made to ensure that degradable and undegradable protein requirements are balance for the target ruminant animal.

B.2. Energy additives

Cereal grains are another source of carbohydrates. Feeds such as maize and small grains can be added to forage at the time of ensiling. Addition of grain to maize silage is not useful, but adding it to hay crop silage has two benefits. First, adding grain to hay crop silage increases the energy content of the silage. This will reduce the amount of supplemental grain that has to be fed. If silage will be the main or only feed offered, then adding some grain to the forage at ensiling will make it a more complete feed. However, grain mixed with silage prior to ensiling or at feeding is nutritionally equal; therefore, if supplemental grain must be fed anyway, no true benefit is realized. Secondly, adding grain to forage will increase the dry matter content of the silage. Hay crops that are not wilted sufficiently prior to ensiling can cause seepage and result in an undesirable fermentation. Added grain may also make wet silage easier to unload from the silo. Recommended application rate for grain is 100-200 lbs./wet ton. This rate will increase the dry matter content of silage by about 5 percentage units. The grain should be cracked prior to ensiling for maximum benefit. In general, adding grain does not improve fermentation because starch (the main carbohydrate in grain) is not readily fermented in the silo. These materials, when added at 150 to 200 pounds per ton of wet forage, will also reduce the moisture content of the chopped forage approximately 5 percentage unit. Commercial products are also available that contain one or more of the above materials.

B.3. Minerals

Minerals such as calcium, phosphorous, sulfur and magnesium have been added to forage at the time of ensiling. Usually these either have no effect on fermentation or act as buffers resulting in higher pH silage. The only reason for adding minerals at the time of ensiling is if the silage will be the only feed offered to the animals. Addition of minerals will make the silage more nutritionally complete. If concentrates are going to be supplemented, it is better to add the minerals to the concentrate mix. Limestone (calcium carbonate) and $MgSO_4$ to increase the calcium and magnesium contents is sometimes added @ 10 to 20 pounds per ton of maize silage to increase the calcium content and extend the fermentation process.

C. Fermentation Inhibitors

Fermentation inhibitors could be used for all types of forages. However, the inhibitors are generally used in wet crops with a low WSC content and/or high buffer capacity (McDonald *et al.*, 1991). A large number of chemical compounds have been tested as potential fermentation inhibitors, but very few have been accepted for commercial use. In the Netherlands, salts of acids have become the most popular fermentation inhibitors (Hogenkamp, 1999). An advantage of these salts is that they are easier and safer to handle than the corresponding acids.

Silage additives inhibiting silage fermentation can reduce clostridial spore counts. In wilted grass silages, a decrease in spore counts by a factor 5 to 20 has been observed. To inhibit clostridial growth, the most effective fermentation inhibitors appear to be additives based on formic acid, hexamethylene and nitrite (Hengeveld, 1983; Jonsson *et al.*, 1990; Lattemae and Lingvall, 1996).

C.1. Inorganic acids and the AIV process

A number of chemical compounds have been tested as potential fermentation inhibitors. Acids are added to forage at ensiling to cause an immediate drop in pH, increase bunk life and limit fermentation losses of protein and carbohydrates. One of the earliest was mixture of mineral acids proposed by A.I.Virtanen, the technique being referred to as A.I.V process (Virtanen. 1933). Usually hydrochloric acid and sulphuric acid were added to the herbage during ensiling in sufficient quantity to lower the pH value below 4.0. These acids are extremely caustic and hazardous to use. This process is very efficient method of conserving nutrients, if carried out properly.

C.2. Organic acid

To inhibit aerobic spoilage, spoilage organisms (i.e. yeasts and acetic acid bacteria) have to be inhibited in their activity and growth. Some additives that have proven to be effective in this respect include chemical additives such as propionic acid, acetic acid, formic acid, Sorbic acid and benzoic acid (Woolford, 1975a; McDonald *et al.*, 1991; Phillip and Fellner, 1992; Weinberg and Muck, 1996). The major benefit of adding weak acids to silage appears to be in reducing spoilage in open storage structures.

C.2.a. Propionic acid

Of the short-chain fatty acids, propionic acid has the greatest antimycotic activity. It is effective in reducing yeast and molds which are responsible for aerobic deterioration in silages. The antimycotic effect of propionic acid is enhanced as pH declines, making it an ideal candidate for improving the aerobic stability of maize silage where pH is low. In the past, aerobic stability was improved when large amounts of propionic acid (1 to 2% of the DM) were added to silage, but the high percentage of acid restricted fermentation in these cases. The application rate of propionic acid additives has varied depending on moisture content of the forage, length of storage and formulation with other preservatives. For example, for high moisture maize with a moisture content of 20% the application rate should be 0.1 and 0.5% for storage of 1 and 6 months, respectively, while it should be increased to 0.8 and 1.1% for 30% moisture of silage for the corresponding lengths of storage. Application rates of 1.5 to 2.0% for haylage and 2.0 to 2.5% for haylage with less than 30% of DM have been suggested. For maize silage, propionic acid at usage rates of 0.2 to 0.5% have been shown to be effective (Beck, 1975). Propionic acid is also difficult to handle because it is corrosive. Thus, salts of acids, e.g., calcium, sodium and ammonium propionate is the most popular fermentation inhibitors (Hogenkamp, 1999). The efficacy of propionic acid and its salts is closely related to their solubility in water. The stronger the bond in between the acid and base, the less soluble the product is and thereby less effective in inhibiting fungi. Among these salts, ammonium propionate is most soluble in water (90%), followed by sodium propionate (25%) and calcium propionate (5%). An advantage of these salts is that they are easier and safer to handle than the corresponding acids. Propionic acid is added at the rate of 0.5 to 1.0 percent of the wet forage weight. The major benefit of adding weak

Table. Categories of silage additives

Class	Subclass	Mode of action	Examples
Fermentation stimulants	Microbial Inoculant	To establish a dominance of efficient lactic acid bacteria	Lactobaccilli
	Fermentable Substrate	Improve lactic acid production To encourage fermentation by provision of fermentable material	Molasses
	Enzymes	To increase fermentation by provision of fermentable material	Cellulase
Fermentation Inhibitors	Direct & Indirect sterilants	To inhibit the microflora either immediately or by sequential release of active agent	Formaldehyde Hexamine
Direct acidifiers	Inorganic acid Organic acids	To lower pH of silage and induce qualitative changes in the microflora	Sulphuric acid Hydrochloric acid Formic acid Acrylic acid
Nutrients	Energy	To improve the nutritional value of silage	Cereals
	Nitrogen	To improve the nutritional value of silage	Urea, Ammonia
	Minerals	To improve the nutritional value of silage	Calcium carbonate
Specific antimicrobial	Antibiotic and other antimicrobial agent	To discourage the growth of spoilage microorganism directly	Bacitracin Streptomycin Bronopol
Aerobic deterioration inhibitors	Organic acid	Inhibit the growth of clostridia	Propionic acid Benzoic acid Sorbic acid

acids to silage appears to be in reducing spoilage in open storage structures.

II b. Formic acid

Formic acid is widely used to restrict fermentation by artificially reducing pH to below 4.0. Formic acid is added to silages at 0.45 percent of the wet weight or 2.25 percent of the dry matter weight (Woolford, 1975b; McDonald *et. al.*, 1991). 0.5% formic acid treated elephant grass had not only an improved fermentation but also higher intake and digestibility compared to the untreated control (Silveira *et al.*, 1980). King grass (*Pennisetum purpureum* × *Pennisetum typhoides*) silage treated with formic acid (3.5 l/t) showed better fermentation quality than benzoic acid treated and untreated silages (Ojeda and Cáceres, 1984). Commercial product most commonly used contains 85 percent formic acid and is applied to the herbage from a container attached to the forage harvester. Complete inhibition of microbial growth does not take place, some lactic acid fermentation occurring. To inhibit the lactic acid bacteria, an application rate of two to three times the normal commercial rate is required.

II c. Acetic acid

Acetic acid also inhibits aerobic spoilage. Acetic acid-type preservative is sodium diacetate. This is mixture of acetic acid and its sodium salt. Recommended rate of application is 1-2 lbs. active ingredient/ton of wet silage. This compound is quite effective in reducing top spoilage. It was established that at pH 5, acetic acid @ 20% of dry matter content of crop would inhibit endospore forming bacteria.

II d. Lactic acid

Lactic acid is fairly effective inhibitor of the endospore forming bacteria, particularly at pH 5. Fermentation inhibitors could in theory may be used for all types of forages. However, in practice they are generally used in wet crops with a low water soluble cabohydrates content and/or high buffer capacity (McDonald *et al.*, 1991). To inhibit clostridial growth, the most effective fermentation inhibitors appear to be additives based on formic acid, hexamethylene and nitrite (Hengeveld, 1983; Corporaal *et al.*, 1989; Jonsson *et al.*, 1990; Lattemae and Lingvall, 1996). Silage additives inhibiting silage fermentation can reduce clostridial spore counts.

III. Formaldehyde

Formalin (35-40% formaldehyde solution) has also been used as a silage preservative, especially aiming at reduced protein degradation in the silo and thus increasing undegradable protein in the rumen of silage-fed animals. Formaldehyde restricts considerably fermentation of silage: 0.8% formalin (w/w fresh basis) almost sterilized an ensiled mass of elephant grass and reduced digestibility of silage (Boin, 1975). A dosage of 0.5% formalin (w/w) applied to a mixture of elephant grass with cassava tops (20.3% DM, 8.5% WSC) reduced ammonia-N and increased precipitable protein in silage, however without suppressing clostridial fermentation (Zanotelli and Mühlbach, 1989). Studies with a 70% formalin plus 26% formic acid plus 4% water mixture applied 0.2% (w/w) to elephant grass (13% DM), aiming at a rate of 4 g formaldehyde/100 g CP in the forage, resulted in poor fermentation quality and impairment of the nutritive value of silages produced (Lavezzo *et al.*, 1984). Accurate formaldehyde rates necessary to improve fermentation in the silo as well as to obtain a protein protection effect in the rumen are difficult to achieve, especially under farm-scale conditions (Mühlbach and Kaufmann, 1979). Thomas (1982) showed that silage treated with formaldehyde improved the utilisation of silage protein

References

Beck, T. 1975. The effect of ensiling agents on aerobic deterioration. *Das Wirtschaft Futter*. 4:8-12.

Boin, C. 1975. Elephant (Napier) grass silage production: effect of additives on chemical composition, nutritive value and animal performance. PhD Thesis, Cornell University.

Bolsen, K. K., Bonilla, D. R., Huck, G. L., Young, M. A., Hart-Thakur, R. A. and Joyeaux, A. 1996. Effect of a propionic acid bacterial inoculant on fermentation and aerobic stability of whole -plant corn silage. *J. Anim. Sci.* 74(Suppl. 1): 274.

Britt, D. G and Huber, J. T. 1975. Fungal growth during fermentation and refermentation of non-protein nitrogen treated corn silage. *J. Dairy Sci.* 58: 1666-1671.

Corporaal, J., van Schooten, H.A., and Spoelstra, S.F. 1989. Invloed van toevoegmiddelen op de kwaliteit van slecht voorgedroogd kuilvoer. Proefstation voor de Rundveehouderij, schapenhouderij en paardenhouderij, Lelystad, The Netherlands. PR Rapport, No.119.

Dawson, T. E. 1994. Propionic acid-producing bacteria as bioinoculants for the preservation of ensiled high-moisture corn. Ph.D. Diss. Michigan State Univ., East Lansing.

Dreihuis, F., S. F. Spoelstra, S. C. J. Cole, and R. Morgan. 1996. Improving aerobic stability by inoculation with Lactobacillus buchneri. Proc. of the XI Intl. Silage Conf., IGER, Aberystwyth. pp. 106-107.

Flores-Galaraza, R. O., B. A. Glatz, C. J. Bern, and L. D. Van Fossen. 1985. Preservation of highmoisture corn by microbial fermentation. *J. Food Protection*. 48:407-411.

Henderson, N. 1993. Silage additives. *Animal Feed Science and Technology*.45: 35-56.

Hengeveld, A.G. 1983. Sporen van boterzuurbacteriën in kuilvoer. Proefstation voor de Rundveehouderij, schapenhouderij en paardenhouderij, Lelystad, The Netherlands. PR Report No. 88.

Huber, J. T., Foldager, J and Smith, N. E. 1979. Nitrogen distribution in corn silage treated with varying levels of ammonia. *J. Anim. Sci.* 48: 1509-1516.

Johnson, C. O. L. E., Huber, J. T. and Bergen, W. G. 1982. Influence of ammonia treatment and time of ensiling on proteolysis in corn silage. *J. Dairy Sci.* 65: 1740-1747.

Jonsson, A., Lindberg, H., Sundas, S., Lingvall, P., and Lindgren, S. 1990. Effect of additives on quality of big-bale silage. *Anim. Feed Sci. Technol.*, 31: 139-155.

Kaiser, E., and Weiss, K. 1997. Fermentation process during the ensiling of green forage low in nitrate. 2. Fermentation process after supplementation of nitrate, nitrite, lactic-acid bacteria and formic acid. *Arch. Anim. Nutr.*, 50: 187-200.

Keady, T.W.J. 1996. A review of the effects of molasses treatment of unwilted grass at ensiling on silage fermentation, digestibility and intake, and on animal performance. *Irish J. Agric. Fd. Res.* 35:141-150.

Kung, L. Jr. and Muck, R.E. 1997. Animal responses to silage additives. In: Silage: Field to Feedbunk, Proceedings from the Silage: Field to Feedbunk. North American Conference, Hershey, PA, pp. 200-210

Lattemae, P. and Lingvall, P. 1996. Effect of hexamine and sodium nitrite in combination with sodium benzoate and sodium propionate on fermentation and storage stability of wilted and long cut grass silage. *Swed. J. Agr. Res.*, 26: 135-146.

Lavezzo, W., Lavezzo, O.E.N.M., & Silveira, A.C. 1984. Efeitos do emurchecimento, formol e ácido fórmico sobre o consumo e digestibilidade de silagens de capim-elefante (Pennisetum purpureum Schum.) *Rev. Soc. Bras. Zoot.*, 13(4): 501-508.

McDonald, P., Henderson, A.R., and Heron, S.J.E. 1991. *The Biochemistry of Silage*. 2nd ed. Chalcombe Publications, Marlow, UK.

Muck, R. E., and L. Kung, Jr. 1997. Effects of silage additives on ensiling. Proc. form the Silage: Field to Feedbunk North American Conference. NRAES -99. pp 187-199.

Mühlbach, P.R.F., and Kaufmann, W. 1979. Eiweiss-Verfuegbarkeit von Formaldehyd-behandelter Grassilage beim Wiederkaeuer. *Das wirtschaftseigene Futter*. 25(2/3): 115-132.

Mühlbach, P.R.F. 2000. Additives to improve the silage making process with tropical forages. In: Silage Making in the Tropics with Particular Emphasis on Smallholders. Mannetje. L.T (ed.). Proceedings of the FAO Electronic Conference on Tropical Silage. FAO, Rome.

Oude Elferink, S.J.W.H., Driehuis, F., Krooneman, J., Gottschal, J.C., and Spoelstra, S.F. 1999. Lactobacillus buchneri can improve the aerobic stability of silage via a novel fermentation pathway, the anaerobic degradation of lactic acid to acetic acid and 1,2-propanediol. p. 266-267.

Pahlow, G., and Weissbach, F. 1996. Effect of numbers of epiphytic lactic acid bacteria and of inoculation on the rate of pH-decline in direct cut and wilted grass silages. XI International Conference, Aberystwyth. pp 104-105.

Phillip, L.E and Fellner, V. 1992. Effects of bacterial inoculation of high-moisture ear corn on its aerobic stability, digestion, and utilization for growth by beef steers. *J. Anim. Sci.*, 70: 3178-3187.

Ranjit, N.K., Cohen, M. A. Smoot, R. C., Tavares, J. Y and L. Kung, Jr. 1998. The effects of Lactobacillus plantarum (LP), L. buchneri (LB) and a propionic acid-based preservative on the fermentation and aerobic stability of corn silage and the aerobic stability of a TMR. *J. Dairy Sci.* 81(Suppl. 1):196.

Silveira, A.C., Lavezzo, W and Silveira, F.S. 1980. Consumo de silagens de capim-elefante (Pennisetum purpureum, Schum.) submetida a diferentes tratamentos. *Rev. Soc. Bras. Zoot,* 9(2): 306-320.

Spoelstra, S.F. 1985. Nitrate in silage. A review. *Grass For. Sci.*, 40: 1-11.

Spoelstra, S.F. 1987. Degradation of nitrate by enterobacteria during silage fermentation of grass. *Neth. J. Agr. Sci.*, 35: 43-54.

Staudacher, W., Pahlow, G., and Honig, H. 1999. Certification of silage additives in Germany by DLG. 12th International Silage Conference, Uppasala. pp 239-240.

Thomas P.C. 1982. Utilization of conserved forages. In: Thompson D.J., Beever De E. and Gunn R.G. (eds), Forage protein in animal production. Brit. Soc. Anim. Prod. Occ. Publ. No. 6. pp. 67-76.

Tosi, H., Rodrigues, L.R. de A., and Jobim, C.C. 1995. Ensilagem do capim-elefante cv. Mott sob diferentes tratamentos. *Rev. Soc. Bras. Zoot.*, 24(5): 909-916.

Virtanen, A.I. 1933. The AIV method of processing fresh fodder. *Empere Journal of Experimental Agriculture.*1: 143-155.

Weinberg, Z.G., and Muck, R.E. 1996. New trends and opportunities in the development and use of inoculants for silage. *FEMS Microbiol. Rev.*, 19: 53-68

Weissbach, F., and Honig, H. 1996. Über die Vorhersage und Steuerung des Garungsverlaufs bei der Silierung von Grunfutter aus extensivem Anbau. *Landbauforschung Volkenrode,* 1: 10-17, Germany.

Whittenbury, R. 1961. An investigation of the lactic acid bacteria. Ph.D. Thesis, University of Edinburgh.

Woolford, M.K. 1975a. Microbiological screening of the straight chain fatty acids (C_1-C_{12}) as potential silage additives. *J. Sci. Food Agr.*, 26: 219-228.

Woolford, M.K. 1975b. Microbiological screening of food preservatives, cold sterilants and specific antimicrobial agents as potential silage additives. *J. Sci. Food Agr.*, 26: 229-237.

CHAPTER 8

Evaluation of Silage Quality

Quality of silage is determined by nutrients content and intake potential of the silage. High quality silage is a stable feed, made from high quality pasture, which is preserved in the absence of oxygen by a high quality fermentation, to minimize any loss of feeding value. It is not possible to produce high quality silage from low quality pasture, no matter how good the fermentation is. The silage quality is generally used to indicate the success of fermentation and not to evaluate the feeding value of the silage. However, the fermentation quality and nutritional value of the silage are usually highly correlated because a poor fermentation can result in great loss of the digestible nutrients (Sutoh *et al.*, 1970; Uchida and Sutoh, 1973). Therefore, both the quality of the ensiled pasture and the quality of the fermentation must be considered.

The quality of the ensiled pasture

The original pasture must be leafy, highly digestible, and contain high concentrations of both metabolisable energy and crude protein. Stage of maturity influences a forage's nutritive value and thus quality of silage. Leguminous forages at budding stage have optimal energy, protein and carotene contents, but DM yield and nutritive value decrease with maturity. Grass at heading stage is highest in caloric value and protein content. When leguminous forages reach at late flowering stage or grasses reach at seed stage, the nutritive value decreases.

The quality of the fermentation

The quality of the fermentation tells us how well the quality of the original pasture has been preserved in the silage. While feeding value of silage is primarily determined by digestibility, it is nonetheless influenced by silage fermentation (Steen *et al*, 1998). Thus, an accurate assessment of silage quality should include a comprehensive range of chemical analyses. Quality of silage fermentation is influenced by several factors, including moisture, water soluble carbohydrate content of raw materials, degree of compaction and effectiveness of final sealing. With well preserved silage, losses in feeding value during fermentation will be small, and the final silage will be only slightly lower in feeding value than the original pasture.

Sampling of silages

It is extremely important that a representative sample of silage be obtained for quality determinations. How useful the analysis of silage will be determined in large measure by analysis of silage as a whole. For chemical fermentation analyses, samples should be frozen immediately after being taken. Sample each field and crop type separately. Individual samples from one field or crop type should be well mixed and a sub-sample taken for analysis. The container used should be sealed to avoid moisture loss and frozen until it is submitted to a laboratory. If samples have not been taken at the time of harvest, take samples as the silage is being removed from the silo. Take care to get individual samples from as many sites as possible from the face of the exposed silage and then combine them into one sample for analysis. It is also important that these samples be taken as soon as possible after the surface of the silage is exposed.

Visual Evaluation

Although silage quality can be estimated by visually examining the silage, but visual examination will not provide accurate information on nutrient content of silage. However, coupled with a chemical analysis, colour, odour and general appearance provide a good indication of the expected overall nutritive value.

Feed quality

High quality silage has lots of leaf, and very little stem in it. Higher the proportion of stem in silage, the poorer is its quality.

Colour

The ensiled material should look similar to or slightly darker in colour than the material before ensiling. Well-preserved silages are green, yellow, or pale brown. A very dark colour or a brown colour indicates that silage is generally poorly preserved and excessive heating caused material to caramelize. This reduces digestibility of the protein. The brown colour in the silage is due to a pigment 'phaecophytin', a magnesium free derivative of chlorophyll. If molds are present in the material, this is an indication that insufficient lactic acid was produced.

Smell

Well preserved silage has a sweet smell. The characteristics aroma of the silage is due to combination of acids and alcohol produced in the silage.

Putrid or rancid smells indicate the presence of butyric acid from poor preservation due to presence of Clostridium bacteria. Livestock will not eat this material readily, and fresh dairy cows are more likely to experience ketosis when consuming silage containing butyric acid. This occurs in haylages that are ensiled too wet and do not reach a pH below 5 because of insufficient lactic acid production.

Vinegar smell indicates high levels of acetic acid. This can discourage intake by livestock and is due to the wrong bacteria dominating the fermentation.

Burnt smell indicates excessive heating occurred.

Moulds and yeast

Moulds grow where silage has been exposed to air. Some fungi can produce toxins. Appearance of molds indicates that there was insufficient lactic acid production. Yeast is more difficult to see, but it can be detected visually if there is larger amount. Yeast on the surface of the ensiled material indicates that not enough material is removed daily to maintain freshness. Yeast found beneath the surface indicates presence of oxygen due to inadequate packing.

Chemical Analyses

Although silage quality can be estimated by visually examining the silage, it can only be estimated accurately by chemical analysis. The most important analyses obtained from feed testing laboratories include dry matter, pH, crude protein, fibre, calcium and phosphorus. Feed test reports can be evaluated in the same way for silages as they can for other forages. However, if the silage has heated too much, the digestibility of protein and dry matter may be greatly reduced. If heating temperatures in excess of 40°C has occurred, as evidenced by a charred appearance or tobacco-like odor, it is useful to have the laboratory perform an acid detergent insoluble nitrogen analysis. In general, if the acid detergent insoluble nitrogen content of a material is above 0.3 per cent of the dry matter or 15 per cent of the total nitrogen in the forage, it is indicative that excessive heating has occurred. pH and concentration of lactic acid, acetic acid, butyric acid and ammonia are indicator of quality of silage.

Silage pH

The best single indicator of the effect of ensiling on the nutritive value of high moisture silage is pH. The pH of an ensiled sample is a measure of its acidity. The relationship between pH and silage quality can be determined. However pH alone is not a totally accurate monitor of silage fermentation. But in general, the lower the pH, better the quality of silage, since it indicates that a lactic acid type of fermentation has occurred. pH is, however, not a good indicator of quality for silages which contain less than 65 per cent water. In general, legume silages have a higher pH than non-leguminous silages and take longer to ensile because of their higher buffering capacity.

Some common reasons for a high silage pH are as follows

- Dry silage (> 50% DM)
- Silage not fully fermented due to early sampling time relative to harvest, cold weather during harvest, and slow or poor packing
- Legume silages with extremely high ash contents (> 15% of DM) and (or) high protein content (> 23-24% CP)
- Silage with excess ammonia or urea
- Clostridial silages
- Spoiled or moldy silages
- Silages containing manure.

Lactic Acid

The primary fermentation acid resulting from a desirable homofermentation is lactic acid. Lactic acid is the strongest and most abundant acid produced during an ideal fermentation. Moderate levels of lactic acid are an indication of good fermentation. Lactic acid should be the primary acid in good silages. Higher the lactic acid better is the quality of silage. Lactic acid should be at least 65 to 70% of the total silage acids in good silage. This acid is stronger than other acids in silage (acetic, propionic and butyric) and thus usually responsible for most of the drop in silage pH. Ideal silage will usually have 3 times more lactic acid than what comprises volatile fatty acids. Depending upon the crop, levels will range from greater than 1-8%. The lactic acid content of good maize and alfalfa silage ranges from 4 to 6% and from 3 to 8%, respectively on a dry matter (DM) basis. In high moisture maize, lactic acid is normally found in the 1 - 3% range.

Lactic acid is the strongest of all silage acids and its presence will drop pH more effectively than the other volatile fatty acids. Fermentations that produce lactic acid result in the fewest losses of dry matter and energy from the crop during storage.

Some common reasons for low lactic acid content are as follows:

- restricted fermentation due to high DM content (especially legumes and grasses with > 50% DM).
- restricted fermentation due to cold weather.
- silages high in butyric acid (Clostridial silages) are usually low in lactic acid.

Acetic Acid

Acetic acid produced in silage provides silages with their characteristic vinegar odor and taste and is usually the predominant acid produced during fermentation for maintaining aerobic stability. Of the volatile fatty acids (VFA), acetic acid is found in the greatest concentration, usually found at less than 3% in silages. High moisture maize usually has low acetic acid content (less than 1.0%). Anything over 3% suggests inefficient hetero fermentative fermentation. High levels of acetic (> 3 to 4%) in any type of silage are indicators of less than desirable silage fermentation.

Extremely wet silages (< 25% DM), prolonged fermentations (due to high buffering capacity), loose packing, or slow silo filling can result in silages with high concentrations of acetic acid (>3 to 4% of DM). In such silages, energy and DM recovery are probably less than ideal. Silages treated with ammonia also tend to have higher concentrations of acetic acid than untreated silage, because the fermentation is prolonged by the addition of the ammonia that raises pH.

A new microbial inoculant (*Lactobacillus buchneri*) designed for improving the aerobic stability of silages causes higher than normal concentrations of acetic acid in silages. However, production of acetic acid from this organism should not be mistaken for a poor fermentation and feeding treated silages with a high concentration of acetic acid does not appear to cause negative effects on animal intake.

Propionic Acid

Propionic acid produces a sharp sweet smell and taste and is usually lower levels of this acid are produced during fermentation for maintaining aerobic stability. Most silage contains very low concentrations of propionic acid (< 0.2 to 0.3%) unless the silage is very wet (< 25% DM). In silages with more typical concentrations of DM (35 to 45% DM), concentrations of propionic acid may be undetectable. Biological additives that theoretically increase the propionic acid concentration of silage usually contain bacteria from the Propionibacteria family. However, these organisms are usually unable to compete in normal silage environments and are thus, usually ineffective. Chemical additives containing propionic acid are more effective for increasing the concentration of this acid in silages. These additives can range markedly in their percentage of active ingredients but most mainstream products will increase the concentration of propionic acid at ensiling from 0 to about 0.15 to 0.30% (DM basis) if added at 2 to 4 lb per ton of wet (~35% DM) silage.

Butyric Acid

Butyric acid produces a rancid butter smell and taste. Quality silage should be less than 0.1% (Harris and Raymond, 1963). Elevated levels indicate silage deterioration from secondary fermentation, which in the presence of unpalatable nitrogenous end products such as amines and amides, may lead to significant reduction in dry matter intake and energy level of the forage. A high concentration of butyric acid (>0.5% of DM) indicates that the silage poorly preserved and has undergone clostridial fermentation, which is one of the poorest fermentations. Silages high in butyric acid are usually low in energy and have undergone extensive protein degradation resulting in large increases in the soluble protein fraction and losses of dry matter. A silage with high levels of butyric acid could have been caused by ensiling a high protein, low sugar forage, such as alfalfa, too wet (< 30% DM). It can also be caused by clostridia from manure that was applied too close to the harvest date (allow 21 to 28 days if possible). Such silages may also be high in concentrations of soluble proteins and may contain small protein compounds called amines that have sometimes shown to adversely affect animal performance. High concentrations of butyric acid have sometimes induced ketosis in lactating cows and because the energy value of silage is low, intake and production can suffer. As with other poor quality silages, total removal or dilution of the poor silage is advised.

Ethanol

Yeast converts sugars to alcohol in silage. Ethanol produces an alcohol smell. High concentrations of ethanol are usually an indicator of excessive metabolism by yeasts. In maize silage, ethanol is between 1 to 3% of the DM but can sometimes be greater than 6 - 7%. The ethanol content in grass and legume silages is usually less than 1.5%. Silages that have high levels of ethanol have usually undergone an extensive fermentation due to yeasts. Although the energy value of such silages is generally good, alcoholic fermentations result in large losses of nutrients. Such silages will also have a tendency to heat rapidly when exposed to air and bunk life will be short. Silages preserved with acids often have high ethanol levels, have high yeast counts, and are very unstable when exposed to air. Dry matter recovery is usually worse in silages with large numbers of yeasts. These silages are also usually very prone to spoilage when the silage is exposed to air. Extremely high amounts of ethanol (> 3 to 4% of DM) in silages may cause off flavors in milk.

Ammonia

Ammonia concentration is usually expressed as a percentage of the crude protein or nitrogen content. In maize silage, ammonia is normally between 5 to 7 % of the CP but it is usually higher (10 to 15%) in grass and alfalfa silage. Low values for ammonia-N indicate minimal breakdown of protein in the silage, usually because pH has fallen quickly to a low level in the silage. High ammonia indicates poor or extensive fermentation, indicating protein breakdown from proteolytic enzymatic activity contained within the crop. High ammonia-nitrogen forages may also be associated with proteolytic activity from Clostridial secondary fermentation. In general, wetter silages have higher concentrations of ammonia. These silages often have high contents of soluble protein. The use or anhydrous ammonia and urea as forage additives will cause high ammonia levels. Extremely wet silages (< 30% DM) have even higher ammonia concentrations because of the potential for clostridial fermentation. Silages packed too loosely and filled too slowly also tend to have high ammonia concentrations. Often times, silage with high concentrations of ammonia coupled with butyric acid may also have significant concentrations of other undesirable end products, such as amines, that may reduce animal performance.

Aerobic Microbial Counts

Yeast, mold, and Bacillus population counts indicate silage bunk life and are expressed in colony forming units per gram of feedstuff (cfu/gm). Aerobic microorganisms require oxygen from air penetration, indicating aerobic instability.

- *Yeast:* Counts less that 100,000 are desirable. Yeast utilizes lactic acid as substrate in the presence of oxygen, causing the elevation of pH. Dry matter losses occur from the production of carbon dioxide, water, and heat. Yeast activity precedes mold growth.
- *Mold:* Counts less than 100,000 are desirable. When lactic acid levels are diminished by yeast and pH rises above 4.5, the silage environment becomes conducive to mold growth. This results in musty, hot, energy depleted, and unpalatable silage.
- *Bacillus:* Counts less than 100,000 are desirable. Fields may be highly contaminated with this aerobe at harvest. Bacillus is highly thermophilic in the presence of oxygen and is primarily responsible for high bound protein (ADIN) values in silages.

Common end products of silage fermentation and their actions

Item	*Desirable or Undesirable*	*Action (s)*
pH	Desirable	Low pH inhibits bacterial activity.
Lactic acid	Desirable	Inhibits bacterial activity by lowering pH.
Acetic acid	Undesirable	Associated with undesirable fermentations.
	Desirable	Inhibits yeasts responsible for aerobic spoilage.
Butyric acid	Undesirable	Associated with protein degradation, toxin formation, and large losses of DM and energy.
Ethanol	Undesirable	Indicator of undesirable yeast fermentation and high DM losses.
Ammonia	Undesirable	High levels indicate excessive protein breakdown
Acid detergent insoluble nitrogen (ADIN)	Undesirable	High levels indicate heat-damaged protein and low energy content

Flieg value

Flieg's value is a widely used method of grading silage in Germany. In this system of grading, particular points are given for pH value and concentration of lactic acid, acetic acid and butyric acid in silage. A particular silage is assigned scores (Fleig's value) according to its pH and fatty acid content as given in table and is graded according to its total score. Score between 85 and 100 indicate very good quality; 60 and 80 good quality; 55 and 60 moderate quality; 25 and 40 satisfying quality and <20, worthless (Flieg, 1938).

pH	*Points*	*Lactic acid (% of total acid)*	*Points*	*Acetic acid (% of total acid)*	*Points*	*Butyric acid (% of total acid)*	*Points*
3.50-3.79	9	Over 60	20	Under 30	10	0.0-0.1	10
3.80-4.20	10	60-55.1	18	30.1-34	9	0.1-2.5	9
4.20-4.40	9	55-50.1	16	34.1-38	8	2.51-5.0	8
4.41-4.60	8	50-45.1	14	38.1-42	7	5.01-7.5	7
4.61-4.80	7	45-40.1	12	42.1-46	6	7.51-10.0	6
4.81-5.00	6	40-35.1	10	46.1-50	5	10.01-15.0	5
5.01-5.20	5	35-30.1	8	50.1-54	4	15.01-20.0	4
5.21-5.40	4	30-25.1	6	54.1-58	3	20.01-25	3
5.41-5.60	3	25-20.1	4	58.1-62	2	25.01-35	2
5.61-5.80	2	20-15.1	2	62.1-66	1	35.01-45	1
5.51-6.00	1	15 & under	0	Over 66	0	45.01-55	0
						55.01-60	-5
						60.1-65	-10
						65.01& over	-20

Summary of silage quality

Parameters	*Good*	*Intermediate*	*Poor*
pH	Below 4.2	4.6-5.2	over 5.2
Lactic acid (%)	3-14	Variable	variable
Butyric acid (%)	Under 0.2	0.2-0.5	over 0.5
Proportion of total acid (%)			
Lactic acid (%)	over 60	40-60	under 40
Acetic acid (%)	under 25	25-40	over 40
Butyric acid (%)	under 5	5-10	over10
Ammonical Nitrogen (% of total N)	10-15	15-30	over 30
ADIN (% of total N)	under15	15-30	over30
Colour	Bright, light green-yellow orgreen brown depending upon material ensiled	Yellowish green to brown green	Very dark green, blue green, grey, or brown
Smell	Lactic acid odor with no butyric[4] acid odor	Slight butyric acid and ammonia odor	Strong butyric acid, ammonia, odor rancid
Texture	Firm, with softer material not easily rubbed from fibre	Softer material can be separated from fibre	Slimy, soft tissues easily rubbed from fibre, moldy
Fleig's value	60 & above	30-60	Less than 20

References

Bal, M. A., Coors, J. G. and Shaver, R. D. 1997. Impact of the maturity of corn for use as silage in the diets of dairy cows on intake, digestion, and milk production. *J. Dairy Sci.* 80:2497–2503.

Flieg, O. 1938. A key for evaluation of silage samples. *Futterb. Garfutteber.* 1:112-128.

Harris, C.E. and Raymond, W.F. 1963. The effect of ensiling on crop digestibility. *J. Brit. Grassl. Soc.* 18: 204–212.

McCullough, M.1984. Feeding quality silage. *Anim Nutr Health:* 30-35.

Steen, R.W.J., Gordon, F.J., Dawson, L.E.R., park, R.S., Mayne, C.S., Agnew, R.E., Kilpatrick, D.J. and Porter, M.G. 1998. Factors affecting the intake of grass silage by cattle and prediction of silage intake. *Anim. Sci.* 66: 115-127.

Sutoh, H., Uchida, S. and Hiramatsu, N. 1970. Studies on chemical composition and quality of silage. 8. The relation between the quality and feeding value of silage. *Sci. Rep. Fac. Agric. Okayama Univ.* 38:51-58.

Sutoh, H. Uchida, S. and Hiramatsu, N. 1973. Studies on chemical composition and quality of silage. 10. The relation between the quality and digestibility of silage. *Sci. Rep. Fac. Agric. Okayama Univ.* 42: 33-38.

Uchida, S. and Sutoh, H. 1973. Studies on chemical composition and quality of silage. 10. The relation between the quality and digestibility of silage *Sci. Rep. Fac. Agric. Okayama Univ.* 24:33-38.

CHAPTER 9

Factors Affecting Silage Quality

Silage fermentation is a dynamic process involving interactions among the forage, microbial populations and the ensiling environment. A good fermentation process is not only dependent on the type and quality of the forage crop, but also on the harvesting and ensiling technique. There are various factors that influence the fermentation process and ultimately the nutrient composition and quality of ensiled feed. These are as follows:

1. Plant Factors

Forage characteristics at the time of ensiling are the predominant factor which determines the final quality of silage. Factors such as type of forage to be ensiled, maturity, DM content and WSC content of that forage, all influence the ease of ensiling and ultimately the quality of silage that is produced.

Type of Forage

The major objective in silage fermentation is to achieve a stable low pH at which biological activity virtually ceases. In this way preservation is obtained whilst minimizing nutrient losses and avoiding adverse changes in the chemical composition of the material. The final pH of the ensiled by-product depends largely on the carbohydrate content in the original materials. Hence, protein-rich feeds with low energy content are very difficult to ensile successfully, and should be mixed with easily fermentable, energy-rich products, e.g. molasses, banana waste or root crops. In general, cereals are easier to ensile than legumes or grasses, because their buffering capacity is lower, and their WSC content is higher. The WSC in cereals provide a readily available source of energy that is rapidly fermented to lactic acid by anaerobic microorganisms. Consequently, pH decline is usually more rapid, and the final pH is usually lower in cereal silage than in grass or legume silage. Because the buffering capacity of grasses and legumes is higher, it takes more acid production to achieve the same reduction in pH decline as obtained in cereal silages. Thus, even when lactic acid production is similar, the pH of alfalfa silage remains considerably higher than that of cereal silage. Because fermentation is inhibited at these low pH values, cereal silages usually contain more residual WSC than do legume silages. This conservation of WSC in cereals can be advantageous because the WSC is also a readily available source of energy for the animal. However, upon exposure to air, these WSC can also be readily utilized by spoilage microorganisms. As a result, cereal silage is often more prone to aerobic deterioration than legume silage.

Forage Maturity

As forage matures from the vegetative stage into a reproductive stage (i.e., flowering, for alfalfa; heading, for cereals) stems and leaves become more lignified, and the digestibility of these plant components declines. Concurrently, WSC levels decline as more carbohydrate is deposited in the plants as cellulose, hemicellulose and starch. The rate of maturation of a crop is influenced by several factors, including variety, moisture level, temperature, nutrient stress and time of season (Fick *et al.* 1988). Optimal timing of harvest usually encompasses a compromise between DM yield, nutrient yield and, with perennial forages, persistence of the stand. Harvesting at earlier stages can increase nutrient yield, but at a cost of reducing total digestible yield throughout the season as well as stand persistence. With annual forages, persistence is not a concern, so maximum nutrient yield becomes the primary consideration. In vitro studies suggested that DM digestibility differs little among barley silages harvested in the soft-dough to the late-dough stages (Baron *et al.* 1992). As barley matures, DM increases and WSC levels in the plant decrease. Higher DM content renders barley of later maturity more difficult to ensile. Because it is more difficult to pack, exclusion of oxygen is less complete. Furthermore, the lower moisture and WSC concentrations may impede fermentation, and as a result, the final pH of ensiled barley harvested in the late-dough stage is usually higher than that of silage from barley harvested at early-dough. Barley harvested at the late-dough stage contains more starch than when harvested at early-dough, but our findings suggest that this higher starch content does not offset the decline in the digestibility of the vegetative components of the plant.

Forage Dry Matter Content

The dry matter content of silage strongly influences the type of fermentation that takes place in a silo. Losses of dry matter in effluents occur when silage dry matter content is less than 25%. Proper moisture at harvest is also critical for compaction of the silage mass, air exclusion, and to provide sufficient moisture to promote lactic acid fermentation. Too much dry matter makes it difficult to pack and expulse the oxygen from the silage mass. Ensiling at lower than normal moisture may lead to aerobically unstable silage with yeast, mold, and bacillus problems. Low moisture silages often have high levels of heat-damaged protein and are monitored by ADIN as a % of total crude protein. Ensiled material should contain more than 50% moisture so that it is easy to compress it tightly in order to obtain good compaction and to eliminate air. The "ideal" dry matter content of the forage depends on the type of silo (which influences the mode and level of compaction that can be achieved).

Ensiling at higher than normal moisture may lead to prolonged fermentation, excessive protein breakdown, and energy loss. Secondary Clostridial fermentation may occur in silages with high moisture levels leading to undesired high levels of butyric and ammonia nitrogen. Excessive moisture, more than 75%, can also be harmful, leading to an undesirable fermentation in later phases, producing sour silage, which reduces palatability and intake. Water can be added and/or wet and dry feeds can be mixed to get the desired moisture content. Often, wilting green forage for 24- 48 hours to increase dry matter is highly desirable because wilted silage requires less lactic acid production and will stabilize at a higher pH level than silage that has a lower dry matter content (Libensperger and Pill, 1987). This is the case because higher dry matter content increases the concentration of soluble dry matter in the silage, and the resulting increase in osmotic pressure inhibits bacterial growth. Thus, fermentation stops in wilted silage because of the high acidity, high osmotic pressure combination (rather than high acidity alone in silage with lower dry matter content). Thus wilting has a "sparing" effect on the level of sugar and the level of fermentation needed to stabilize the silage. As a result, wilting is particularly important when the ensiled forage is a legume, which has a relatively low level of fermentable sugars as compared to grass or maize. Wilting tends to increase the number of lactic acid bacteria present at ensiling. This improves the likelihood of an early start of lactic acid fermentation and reduces the need for commercial lactic acid bacteria inoculants. Also, wilted silage is usually more palatable, in part because of lower acid content.

In some cases, it is desirable to increase the DM content of forage. High moisture silages (20 to 27% DM) promote a very active fermentation and they are often associated with increased in seepage losses from the silo. Furthermore, intake of high moisture silages also tends to be reduced relative to intake of forage ensiled at optimal (27 to 38%) DM (Demarquilly *et al.* 1977; Thomas *et al.* 1961). DM content of forage tends to increase with advancing maturity of the crop, but silage DM can also be increased by wilting less mature forage in the field prior to ensiling. In this case, the higher DM is achieved without the increased lignification associated with more mature plant cell walls. Thus, wilting can be used as an effective tool to elevate forage DM into an acceptable range for ensiling. Over wilting the forage, however, can reduce silage quality. Loss of DM from forage during wilting can be as high as 4% per day (McDonald *et al.* 1991) and prolonged plant respiration can further reduce WSC levels. Reducing available WSC reduces the amount of lactic acid produced. Consequently, the pH of wilted silage is often higher than silage that is chopped directly. Lower WSC levels may also account for the lower rate of digestion and effective degradability observed with wilted silage as compared to direct-cut silage. Wilting, therefore, can also affect ruminal digestive characteristics as well as nutrient value of silages, although not to as great an extent as forage maturity.

Forage Variety

Although there is no doubt that forage variety can impact in vitro digestibility of silages, this varietal effect has seldom resulted in consistent changes in animal performance (i.e., milk production, growth rate). Neutral detergent fibre (NDF) digestibilities differing by 7 to 8 percentage points have been measured among maize hybrids (Vattikonda and Hunter 1983), although these differences were less dramatic when multiple-year studies were conducted in numerous locations (Allen *et al.* 1993). Barley silages differed in their chemical composition, but not in their in vitro digestibility (Edwards *et al.* 1968), nor did barley variety affect milk production in dairy cows (Kennelly and Khorasani 1999). Assessing the effect of forage variety on animal performance is technically difficult, as it is almost impossible to ensure identical harvest conditions among the varieties being studied. In general, varietal differences are overshadowed by differences in growing conditions, harvest time and ensiling practices. A variety selected for optimal silage production will still result in poor quality silage if it is harvested at improper maturity or it is poorly ensiled. As one cannot control the weather, selection for maximum yield may still be the best bet when trying to select a variety for silage.

Length of chopping

The finer the chopping, the better is the compaction and therefore the more successful the storage, due to the effective exclusion of air. Chopping into small pieces can be done by hand or with a stationary forage chopper.

2. Microbial Populations

Epiphytic Microbial Populations

Epiphytic microbial populations refer to those microorganisms which are associated with the forage prior to ensiling. The survival and activity of these populations is also among the factors influenced by the characteristics of the crop at the time of ensiling. For example, epiphytic microbial populations tend to decline with increasing forage DM and with increased exposure to solar radiation. This may also partially explain why ensiling high (> 40%) DM forages is difficult. It has been proposed that the forage should contain at least 1,000,000 colony forming units (CFU) per gram for successful fermentation (Spoelstra, 1991). The types of lactic acid bacteria present in the epiphytic populations, as opposed to the total numbers of bacteria, may be more important in determining the efficiency of the fermentation process. Homofermentative LAB produce only lactic acid as an end product, whereas heterofermentative LAB produce additional end products such as acetic acid. Enhanced production of lactic acid can cause a more rapid decline in silage pH, thereby conserving the nutrients in the silage. This relationship is the foundation for selection of bacterial strains as silage inoculants.

Forage destined for the silo may also harbor undesirable epiphytic microorganisms that can contribute to spoilage and could be a potential health risk. Application of manure onto forage prior to ensiling can increase numbers of epiphytic enterobacteria such as *Bacillus* and *Clostridium* spp. (Rammer *et al.* 1994). Contact of the forage with soil can also increase yeast and mold counts in the silage. Although these microorganisms are usually inactivated during ensiling, they can become active and contribute to accelerated spoilage when the silage is exposed to air upon feeding.

Silage Inoculants

Selection of microbes for inclusion in a silage inoculant is the principal factor that will influence the impact of the product on silage fermentation and subsequently, animal performance. *Lactobacillus plantarum, Lactobacillus acidophilus, Enterococcus faecium, Pediococcus acidilactici* and *Pediococcus pentosaceus* are the most frequently used species, but other species such as *Streptococcus bovis* (Jones *et al.*, 1991), *Streptococcus thermophilus, Serratia rubidaea* (Phillip and Fellner, 1992) and *Lactobacillus buchneri* (Weinberg and Muck, 1996) have been considered. Many recently developed inoculants contain multiple species, building on evidence that growth of one bacterial species may facilitate growth of another (Fitzsimons *et al.*, 1992). For example, *Pedicococcus* spp. and *Enterococcus* spp. grow more rapidly and are more tolerant of high DM conditions than are *Lactobacillus* spp. Rapid growth of *Pedicococcus* and *Enterococcus* increases the rate of acid production in recently ensiled forage, and the rapid decline in pH facilitates the establishment of lactobacilli as predominant microorganisms in the fermentation process. This synergism has only been demonstrated experimentally with combinations of two or three species, however, other than as a marketing feature, there is likely little benefit to including numerous species in a product. Selecting particular strains of a single species for inclusion in an inoculant is likely just as important as deciding what combination of species to include. Several studies have shown that fermentation responses differ widely among strains of the same species (Woolford and Sawczyc, 1984; Hill, 1989; Fitzsimons *et al.*, 1992). Thus, two products containing identical bacterial species, but different strains, could exert different effects on silage fermentation and, consequently, animal performance.

The mechanisms by which inoculants may alter silage fermentation and potentially improve animal performance are numerous. The most frequently observed effects of inoculants are accelerated post ensiling decline in pH and increased numbers of lactobacilli. Water soluble carbohydrates in the silage are conserved by the increased rate of fermentation (Gordon, 1989), and ammonia levels are reduced as a result of suppressed deamination and proteolysis (Heron *et al.*, 1989). Higher lactic acid levels can also raise propionate levels in the rumen. This may give rise to the energetic advantages associated with increased propionate absorption (Van Vuuren *et al.*, 1995). Conservation of protein and WSC is likely more critical with legume and grass silages than with cereal silages, as the former contain higher levels of protein and lower levels of WSC. Inoculants commonly reduce the pH of grass and legume silages, but seldom alter pH in cereal silages, in which the final pH often falls below 4 even without additives. Additional benefits may arise from nutrient conservation due to improved DM recovery (Henderson, 1993) and aerobic stability (McAllister *et al.*, 1995). Maize silage, inoculants improved DM recovery by 1.3% and feed efficiency by 1.8% (Bolsen *et al.*, 1992). The inoculant *L. plantarum* MTD1 improved DM intake by 4.8% and milk production by 4.6% when it was applied to grass, maize or alfalfa (Moran and Owen, 1994). As a result of the inoculants milk production increased by an average of 0.816 kg/d and average daily gain by 11.9% (Kung and Muck, 1997).

Kluyveromyces lactis

When the silo is opened for feeding or the anaerobic environment of the silo is replaced by an aerobic environment, silage undergoes "secondary fermentation", due to the rapid growth of yeasts in the aerobic environment and abundant energy source. Some kinds of yeasts in silage survive under acidic conditions even if lactic acid fermentation proceeds and they can use lactic acid as an energy source in an aerobic environment. By the consumption of lactic acid, the pH of silage increases to the neutral range and the environment becomes favorable for the growth of various kinds of aerobic bacteria. Some kinds of yeast strains produce killer proteins to inhibit the growth of other yeast strains. Therefore, a killer yeast strain, *Kluyveromyces lactis,* with a wide killing spectrum was used to prevent aerobic spoilage of silage by wild yeasts. This yeast displayed a killing activity against wild yeasts, but assimilated lactic acid with possible silage spoilage. Therefore, the gene coding the key enzyme for gluconeogenesis (phosphoenol- pyruvate carboxykinase) was disrupted to construct a killer strain defective in the ability to grow on lactic acid as a sole carbon source. It was concluded that this genetically modified yeast significantly prolonged the aerobic stability of silages in the model system and actually reduced aerobic spoilage of silage (Lowes *et al.*, 2000).

3. Ensiling Environment

Selection of a Silo

Several factors, including type of silo, amount of forage to be ensiled and associated machinery affects silage quality. In general, concrete or oxygen limiting towers are associated with lower DM losses but require higher capital expenditure per tone of forage ensiled. Bunker silos are relatively inexpensive to construct, but suffer from the highest DM losses and are the least forgiving of errors in silage management. Dry matter losses from bagged silage are low, and bagging is an economical method of storing relatively small amounts of silage for short periods of time.

Packing density in a silo and rapid sealing (air-tight)

The gap space between the plant particles left in a silo, when the packing density of the silage materials is low, delays the formation of an anaerobic environment even if the silo is rapidly sealed. As a result, the nutritive value of the materials decreases due to the activity of aerobic microbes and plant respiration. Therefore, high packing density of materials and rapid sealing of the silo are required to enhance anaerobic conditions and promote lactic acid fermentation by LAB. Although plant cells and aerobic microorganisms utilize oxygen after the forage is ensiled, steps must be taken to minimize the penetration of oxygen into the forage from the surrounding atmosphere. In bunker silos, exclusion of oxygen is obtained by packing the silage to an optimal density and sealing the packed silage with an atmospheric barrier such as polyethylene sheets. Packing increases the density of the forage in the silo and consequently reduces porosity and the penetration of air into the silage mass (Ruppel, 1993). As air is excluded, aerobic respiration is precluded and losses in DM are minimized. Also, storage of forage at a higher density increases the capacity of the silo and as a result reduces the capital investment in the storage structure per tonne of ensiled forage.

The density of the silage will depend on the type of implement used during packing as well as the total time spent packing per tonne of forage. Packing density is maximized by using implements with tires that apply the greatest weight per unit of surface area. Several factors including forage delivery rate, number and weight of packing tractors, and total time spent packing can influence the packing time per tonne of forage. Optimally, forage should be packed at a rate of 1 to 4 minutes per tonne of silage. Packing rates lower than 1 min per tonne of forage may indicate that the forage is being delivered to the bunker too quickly, that there are insufficient packing tractors, or that the packing layer thickness is too large. Maintaining a shallow packing layer of 15 cm or less can also be used as management strategy to increase packing density. It is recommended that silage be packed to a minimal density of 225 kg of DM per m3. The improved DM recovery associated with packing to higher densities (e.g., 240 to 260 kg of DM per m3) has to be weighed against the increased labor and machinery costs associated with longer packing times. Because bunker silos have substantial surface area exposed to the atmosphere, penetration of both air and moisture can affect the ensiling process and consequently, the nutritional value of the silage (Bolsen, 1997). Polyethylene sheeting weighted with tires is the most common and effective method of excluding oxygen and moisture. Other surface coverings such as limestone, molasses and sawdust are of limited value. The economic savings of covering bunker silos has been well documented, to the point that it is the "no-brainer" of silage management. Losses of DM in an uncovered silo are extensive. Up to 30% of the DM may be lost from the uppermost 1 m of silage which, in a typical bunker silo, can represent 25% of the total forage in the silo.

Storage Temperature

The storage temperature is one of the important factors for the fermentation change from butyric acid to lactic acid. Although lactic acid fermentation is delayed at 25°C, butyric acid fermentation is strongly inhibited, which is one of the reasons why it is very difficult to make good quality silage in the summer season with temperatures above 30°C.

Aerobic Deterioration

When the silo is opened for feeding or the anaerobic environment of the silo is replaced by an aerobic environment, silage undergoes changes sometimes referred to as "secondary fermentation", due mainly to the rapid growth of yeasts in the aerobic environment and abundant energy source. Some kinds of yeasts in silage survive under acidic conditions even if lactic acid fermentation proceeds and they can use lactic acid as an energy source in an aerobic environment. By the consumption of lactic acid, the pH of silage increases to the neutral range and the environment becomes favorable for the growth of various kinds of aerobic bacteria. As a result, not only the nutrient loss from silage but also a reduction of silage intake by cattle occurs. The majority of silage DM loss takes place after the forage is ensiled, procedures adopted during the feed-out phase represent an important component of any total silage management program. Some silage begins to heat within hours of aerobic exposure, whereas others remain stable for days or even weeks (McDonald *et al..* 1991). Legume silages with low DM and WSC contents are generally more resistant to aerobic deterioration than are cereal silages, in which WSC concentrations are sufficiently high to promote the growth of yeasts and molds (Weinberg *et al.* 1993). Furthermore, legume silages may also contain other unidentified microbial inhibitors that prevent the growth of spoilage microorganisms (Jones *et al.* 1995). It was found that some inoculants can improve the aerobic stability of barley silage by inhibiting the growth of both yeast and molds (McAllister *et al.* 1995). To minimize aerobic deterioration, silos should be designed to match the feed demand rate such that at least 4 to 6 inches of silage are removed from the face on a daily basis. This technique prevents any one location in the silo from being exposed to oxygen for a period of time sufficient to foster proliferation of the aerobic microorganisms responsible for spoilage. Accumulation of loose silage at the bottom of face should also be avoided as this unpacked material is especially vulnerable to rapid aerobic decomposition.

Thus, to make good quality silage, one must have an appreciation of the plant, microbial and environmental factors that influence the fermentation process and ultimately, the nutrient value of the silage. These factors must be considered as an integrated package, as neglect of any one component can lead to a breakdown in this forage preservation process. Silage inoculants can facilitate the ensiling process, but they are not a replacement for paying attention to the fundamental factors (plant maturity, dry matter content, oxygen exclusion) that are the keys to making good quality silage.

References

Allen, M.S., Oba, M. and Choi, B.R. 1993. Nutritionist's perspective on corn hybrids for silage. *In*: Silage: Field to Feedbunk. Proceedings from the Silage: Field to Feedbunk North American Conference, Hershey, Pennsylvania, pp. 25-36.

Baron, V.S., Dick, A.C. and Wolynetz, M.S. 1992. Characterization of barley silage - maturity relationships for central Alberta. *Can. J. Plant Sci.* 72:1009-1020.

Bolsen, K.K. 1997. Issues of top spoilage losses in horizontal silos. In: *Silage: Field to Feedbunk*. Proceedings from the Silage: Field to Feedbunk North American Conference, Hershey, PA, pp. 137-150.

Bolsen, K.K., Huck, G.L., Siefers, M.K., Schmidt, T.E., Pope, R.V. and Uriarte, M.E. 1999. Silage management: five key factors. Kansas State University, Manhattan, KS.

Bolsen, K.K., Sonon, R.N., Dalke, B., Pope, R., Riley, J.G. and Laytimi, A. 1992. Evaluation of inoculant and NPN silage additives. A summary of 26 trials and 65 farm-scale silages. Report of Progress, Kansas State University.

Demarquilly, C and Dulphy, J.P. 1977. Effect of ensiling on feed intake and animal performance. In: Proceedings of International Meeting on Animal Production from Temperate Grasslands, Dublin, pp. 53-61.

Edwards, R.A., Donaldson, E. and MacGregor, A.W. 1968. Ensilage of whole-crop barley. I. Effects of variety and stage of growth. *J. Sci. Food Agric.* 19: 656-660.

Fitzsimons, A., Duffner, F., Curtin, D., Brophy, G., O'Kiely, P. and O'Connell, M. 1992. Assessment of Pediococcus acidilactici as a potential silage inoculant. *Appl. Environ. Microbiol.* 58:3047-3052.

Gordon, F. J. 1989. An evaluation through lactating cows of a bacterial inoculant as an additive for grass silage. *Grass and Forage Sci.* 44:169-179.

Henderson, N. 1993. Silage additives. *Anim. Feed Sci. Technol.* 45:35-56.

Heron, S.J.E., Edwards, R.A. and Phillips, P. 1989. The effect of pH on the activity of ryegrass proteases. *J. Sci. Food Agric.* 46:267-277.

Hill, H. A. 1989. Microbial ecology of lactobacilli in silage. In: Food for Thought - Proceedings of the 2nd Forage Symposium, Pioneer Hi-Bred International, Johnston, IA, pp. 47-64.

Jones, B.A., Muck, R.E. and Hatfield, R.D. 1995. Red clover extracts inhibit legume proteolysis. *J. Sci. Food Agric.* 67:329-333.

Jones, B.A., Muck, R.E. and Ricke, S.C. 1991. Selection and application of Streptococcus bovis as a silage inoculant. *Appl. Environ. Microbiol.* 57: 3000-3005.

Kennelly, J. and Khorasani, R. 1999. Nutritive values and fermentation characteristics of barley silage varieties. In: Silage Production in Relation to Animal Performance, Animal Health, Meat and Milk Quantity. XII International Silage Conference, Uppsala, Sweden, pp. 104-105.

Kung, L. Jr. and Muck, R.E. 1997. Animal responses to silage additives. In: Silage: Field to Feedbunk, Proceedings from the Silage: Field to Feedbunk North American Conference, Hershey, PA, pp. 200-210

Lowes, K, F., Shearman, C.A., Payne, J., Mackenzie, D., Archer, D.B., Merry, R.J and Gasson, M.J. 2000. Prevention of Yeast Spoilage in Feed and Food by the Yeast Mycocin HMK. *Appl. Environ Microbiol.* 66(3): 1066–1076.

McDonald, P., Henderson, N. and Heron, S. 1991. The Biochemistry of Silage. Chalcombe Publications, Marlow, UK.

McAllister, T.A., Selinger, L.B., McMahon, L.R., Bae, H.D., Lysyk, T.J., Oosting, S.J. and Cheng, K.J. 1995. Intake, digestibility and aerobic stability of barley silage inoculated with mixtures of Lactobacillus plantarum and Enterococcus faecium. *Can. J. Anim. Sci.*, 53:171-181.

Moran, J.P. and Owen, T.R. 1994. The effects of Ecosyl-treated silage on milk production by lactating dairy cows. Proceedings of the National Conference on Forage Quality, Evaluation and Utilization. University of Nebraska, Lincoln, NE.

Phillip, L.E. and Fellner, V. 1992. Effects of bacterial inoculation of high-moisture ear corn on its aerobic stability, digestion and utilization for growth of beef steers. *J. Anim. Sci.*, 70:3178-3187.

Rammer, C., Ostling, C., Lingvall, P. and Lindgren, S. 1994. Ensiling of manured crops - effects on fermentation. *Grass Forage Sci.* 49:343-351.

Ruppel, K.A. 1993. Bunker silo management and its effects on haycrop quality. In: Silage Production from Seed to Animal, Proceedings from the National Silage Production Conference, Syracruse, New York. pp. 67-84.

Thomas, J.W., Moore, L.A., Okamoto, M. and Sykes, J.F. 1961. A study of the factors affecting rate of intake of heifers fed silage. *J. Dairy Sci.* 44:1471-1483.

Van Vuuren, A.M., Huhtanen, P. and Dulphy, J.P. 1995. Improving the feeding and health value of ensiled forages. In: Recent Developments in the Nutrition of Herbivores. Eds. M. Journet, E. Grenet, M.-H. Farce, M. Theriez, C.M. Demarquilly. Proceedings of the IV International Symposium on the Nutrition of Herbivores. INRA Editions, Paris, France, pp. 279-307.

Vattikonda, M.R. and Hunter, R.B. 1983. Comparison of grain yield and whole-plant silage production of recommended corn hybrids. *Can. J. Plant Sci.* 63:601-609.

Weinberg, Z.G. and Muck, R.E. 1996. New trends and opportunities in the development and use of inoculants for silage. *FEMS Microbiol. Rev.* 19:53-68.

Weinberg, Z.G., Ashbell, G., Hen, Y. and Azrieli, A. 1993. The effect of applying lactic acid bacteria at ensiling on the aerobic stability of silages. *J. Appl. Bacteriol.* 75:512-518.

Woolford, M.K. and Sawczyc, M.K. 1984. An investigation into the effect of cultures of lactic acid bacteria on fermentation in silage. 1. Strain selection. *Grass Forage Sci.*, 39:139-148.

CHAPTER 10

Silage From By-Products

It is recognized in developed countries that the production of silage of high quality cultivated forage can be a valuable component for the development of a high-performing and low-cost system of animal production, using a relatively low level of purchased concentrates. However, this appears inappropriate for smallholders in tropical countries, primarily due to lack or high cost, or both, of equipment for harvesting and conservation and production of cultivated forage being often limited due to lack of available land. Many agricultural, agro-industrial and fishery by-products have potential as animal feeds. Most farmers in developing countries rely for their food security on the cultivation of cereals, root crops and high-value crops such as fruits and vegetables, which understandably take priority in the allocation of land. Fruit, vegetables and root crops are increasingly integrated in the farming system and play a key role as staples in the human diet in most developing countries.

In tropical and sub-tropical countries, there is a wide range of by-products and residues from food crops and food processing that is potentially valuable feed supplements. Consequently, there is a wide range of valuable by-products and residues resulting from food cropping systems and food processing, which are often inefficiently or totally under-utilized and wasted. Many of these products currently are completely unused or are largely wasted due to the inability of farmers to use them before they spoil, as a result of seasonal production peaks and troughs. Consequently these by-products often become pollutants. The ensiling of by-products is a simple and appropriate method of conservation (Lien *et al.*, 1994). It is the most effective way to improve animal feed resources through the rational use of locally available agricultural and industrial-by products likely to be available to small-scale farmers at village level (Chedly and Lee, 1999). Ensiling can also render some previously unpalatable products useful to livestock by changing the chemical nature of the feed. Silage is a very versatile product and can be used as a basal diet or as a concentrate type supplement to other feeds such as forage. The most common by-products from various root and tuber crops, fruit crops, agro-industry and animal industries are discussed in the following sections.

By-Products Suitable to Silage Making

In tropical and sub-tropical countries there is a wide range of by-products and residues from food crops and food processing that is potentially valuable feed supplements, without considering residues from cereal grain. The most common by-products from various root and tuber crops, fruit crops, agro-industry and animal industries are the following:

1. **Brewer's spent grains**

The extracted malt or spent grain contains 75-80% water when filtered off. Wet spent grain spoils quickly and should be used fresh or stored out of contact with air. It can be stored for up to two weeks quite successfully by heaping it, treading it well and covering with sacks or plastic sheets. For longer storage, it may be ensiled in an airtight trench silo with drainage. Ensiling in tightly tied plastic bags is also an effective storage method. Wet spent grain can be ensiled alone or in association with other ingredients, such as with 2-3% molasses (to ensure proper fermentation), chopped banana by-products (trunk, pseudostems, fruit, peel) or chopped cassava. The latter has the advantage of absorbing the juice from spent grain and consequently limiting losses during fermentation. The quantities to be incorporated depend on availability. However, in any case, the DM content in the ensiled mass should not exceed 45%, in order to ensure proper fermentation.

Brewer's spent grain is a valuable potential supplementary feed for livestock. It is a safe feed when it is used fresh or properly stored. It is a relatively bulky feed, but a good source of energy and protein. It can be used to feed beef cattle (10-15 kg daily) and calves (2-4 kg daily).

However, it is far more suitable in rations for dairy cows. Spent grain is a balanced feed for dairy cows and has a good reputation for stimulating milk production. Milk yield response is very rapid and production significantly increases when lactating cows are fed spent grain as a supplement. Wet grain can be given in large amounts to dairy cows, up to 15 kg per day. In order to avoid off-flavours in the milk, it is recommended that spent grain be fed to cows after rather than before milking. When it is eaten in large quantities (15-20 kg/day), distribution of 100-150 g of sodium bicarbonate given twice daily is recommended in order to prevent rumen acidosis disorder.

2. **Banana by-products**

Bananas are grown in almost all farms in the humid tropics and constitute one of the staple foods for human consumption. Banana plant waste is of considerable importance in feeding ruminants. The by-products are the following:

Rejected banana

Rejected bananas - green, immature and ripe - are a good source of energy supplement for grazing or stall fed animals. Dairy cows relish them and can consume them in fairly large amounts. However, bananas have low contents of Crude Fibre, Crude Protein and minerals, and should therefore be fed together with grass or another source of roughage (in order to avoid rumen disturbance) as well as with protein and mineral supplementation. When the rejected bananas are widely available, good silage can be made from chopped bananas mixed with one or several protein-rich feeds, such as poultry litter, dry spent grain, fish waste and cassava leaves.

Banana leaves and pseudostems (trunks)

Banana pseudostems (trunks) and leaves are useful sources of roughage in many tropical countries, mainly during the dry season. They can be chopped and fed fresh or ensiled. As pseudostems are low in protein and minerals, they are more efficiently used when supplemented with rich-protein ingredients, such as copra meal, multi-nutrient feed blocks, cassava leaves, poultry manure or spent grain.

The use of chopped and ensiled pseudostems is particularly recommended when the bunch has been harvested and plants are cut down; the large quantity of trunks available at harvest time can be safely preserved through a well-planned silage operation. The silage is of good quality when chopped pseudostems are properly mixed with an easily fermentable carbohydrate (such as molasses or sliced root vegetables) and protein-rich feeds (such as poultry litter or wet spent grain).

3. **Root crops**

The major root crops that are potentially available for animal feeding in tropical countries are cassava, taro, sweet potato and yams.

Cassava roots

Fresh and sun-dried cassava roots are consumed by ruminants in different forms (sliced, chopped, ground) and used as a substitute for cereal grains in many countries. Cassava roots are a good energy source for dairy cattle, because they are high in carbohydrate, which is an important and readily available energy source for rumen microbes. However, they are low in protein. They are particularly efficiently used by high producing cows and in the first stage of lactation. Cassava roots can be given in large quantities: up to 25% of total DM intake. However, protein and mineral supplements must be fed in order to balance the ration. As cassava roots are rich in easily fermentable carbohydrates, they constitute an excellent energy additive when they are chopped and ensiled with other feed resources, such as fish waste, cassava leaves, banana pseudostems, spent grain, poultry litter, etc.

Cassava Leaves

The cassava leaves are also a potential and valuable protein feed for ruminants. It is estimated that when leaves are harvested at the same time as the roots, yields are in the range of 1 to 4 t DM/ha. Fresh cassava leaves have been successfully fed to cattle, including dairy cows, in many countries. Fresh raw cassava contains cyanogenic glucosides (HCN compounds) that are toxic to monogastrics, but leaves can be fed fresh to ruminants, as rumen micro-organisms appear to be able to detoxify the HCN. However, this is easily removed during processing by sun drying or ensiling. Ensiling cassava leaves is the simplest method, not only to significantly reduce HCN concentrations to safe levels for monogastrics, but also to preserve the nutritive value of harvested cassava leaves for efficient use for off-season feeding of dairy cows. The freshly harvested leaves are first chopped, and can be ensiled alone or mixed with energy-rich feeds, such as banana wastes, root vegetables, wet spent grain, etc. The whole cassava plant (including root and aerial parts) can also be chopped and ensiled in the same way. The silage is fairly balanced for dairy cows.

Sweet potato

Sweet potato is another root crop grown by farmers in many tropical countries. The by-products are roots, leaves and vines. Roots have low protein, fat and fibre concentrations, but they are high in carbohydrates, whilst foliage has a lower carbohydrate concentration but higher fibre and protein concentrations. The vines, which are usually wasted, can serve as a nutritive and relished green supplement for cattle. A mixture of waste bananas, cassava roots and sweet potato tubers and leaves can be ensiled effectively without the need for additives.

Yams

Yams are also widely grown in many tropical countries. Their nutritional value is limited by their bitter alkaloid concentration, as well as of tannins and saponins. Yams must be cooked to improve their nutritional usefulness if they are to be fed to monogastrics or calves. The by-products include roots, offcuts, leaves and vines. Vines are a valuable cattle feed and can be successfully ensiled mixed with other feed ingredients indicated above.

Taro

Taro is the staple food of the population in many tropical countries. Taro by-products include roots, trimmings, leaves and stems: all potentially valuable feed supplements. Taro roots are outstanding as a feed, being particularly rich in energy. Raw taro contains substances that irritate the tongue and palate of animals, so that it must be cooked to improve its nutritional usefulness, mainly for monogastrics. The leaves are rich in protein and are relished by cattle. Taro by-products can be chopped and ensiled in association with the aforementioned feed resources. Silage making reduces considerably undesirable substances in taro by-products, which thus become more appetising.

4. Wet pulps from fruit and vegetables

Fruit wastes

Fruit wastes (like mango, papaya, pineapple, citrus, etc.) and leaves are potential feed resources. The most suitable method for conserving these materials is to ensile them with the aforementioned ingredients, to ensure good fermentation and enhance the silage quality with their high sugar concentration. Citrus pulp is the residue remaining after the extraction of citrus juice: it represents approximately half of the fruit and has a mean DM content of 20% (Boucqué and Fiems, 1988). Citrus pulp is an energy-rich feed resource with high metabolizable energy content. The ensiling of the citrus pulp, damaged fruits and leftovers, in combination with other by-products with high protein concentration, such as poultry litter and wheat bran, is a simple and appropriate method of conservation that can make a significant contribution to the feeding requirements of ruminants, mainly for high-yielding cows in early stages of lactation. In addition, ensiling citrus wastes has advantages over traditional drying, in that less energy is used, cost of processing is much reduced and there is improved palatability (AI-Rokoyan *et al.*, 1998, Ashbell, 1992).

Tomato pulp

Tomato is the most popular vegetable crop grown in Mediterranean countries where there are numerous tomato-processing plants. Tomato processing residues called tomato pulp (a mixture of peel and seed) accounts for about one-fifth of fresh weight and has a high nutritional value. It is a particularly protein-rich feed resource. Fresh tomato pulp becomes sour and mouldy rapidly because it is traditionally processed during summer and has high moisture content, approximately 80-84%. Consequently, it is advisable to ensile tomato pulp in alternate layers with dry by-products, such as chopped straw, wheat bran and poultry litter, so that the liquid effluent is absorbed and used.

Pressed olive cake

The production of olive oil is basically confined to the Mediterranean basin. Large amounts of crude olive cake, a residue of kernels and pulp left after oil extraction, are produced yearly. Despite its low nutritive value, this by-product is a potential feed resource, mainly during periods of feed shortage. Usually, crude olive cake stored in heaps next to the processing plant deteriorates quickly because of its high lipid concentration - between 10 and 14%. The voluntary intake of such cake decreases with storage duration. Ensiling fresh olive cake either alone or with high-quality by-products, such as wheat bran, poultry litter or tomato pulp, improves storage quality and gives a well-preserved, palatable feedstuff. This technique was tried ten years ago in Tunisia (Kayouli *et al.*, 1993), and then successfully adopted by many smallholders in the vicinity of olive oil processing plants.

Grape marc

After pressing 100 kg of grapes, there remains 5 to 10 kg of grape marc (seed and pulp), with about 50% DM content and relatively low nutritive value. Ensiling fresh grape marc either alone or with high-quality by-products, such as wheat bran, poultry litter or tomato pulp, improves storage quality and gives a well-preserved, palatable feedstuff (Hadjipanayiotou, 1987).

5. **Fish by-products**

Fish by-products are usually obtained from inedible whole fish or from waste from fish processing industries. This is an excellent source of protein and minerals for livestock, mainly for cows that have recently calved and for high-yielding cows. The ensiling of the by-products in combination with other easily available feeds rich in fermentable carbohydrates, such as molasses, sweet potato or cassava roots, is a simple and appropriate method of conservation that has been successfully applied recently in some countries. In all cases, the maximum amount of fish wastes that can be included in the silage is 50% with a dry source of carbohydrates, and much lower - about 10% - with fresh sources.

6. **Poultry litter**

Animal wastes such as poltry manure, pig excreta and cow waste can be ensiled along with green fodder. Poultry litter and manure contain about 25% crude protein on a DM basis, about half of which derives from uric acid, which can be efficiently used by rumen microbes for protein production. Similarly pig exreta contains 15-20% crude protein and ruminant wastes contain 10-16% crude protein. Poultry litter is also rich in minerals. The results of many experiments around the world indicate that dried or ensiled poultry litter can be successfully included in the feed of ruminants as a protein supplement. The ensiling of the poultry litter is a simple and appropriate method of conservation. The silage prepared with the incorporation of animal wastes has been named 'wastelage'. The process of ensiling animal wastes to be ensiled has some limitations such as: high buffering capacity of animals waste due to the presence of high level of ionic compounds and high microbial load (Kamra and Pathak, 1997). Threfore, ensling process must be sufficiently efficient so as to bring about these changes n animal wastes and make it fit for feeding to animals.

A good quality silage was prepared by fermentating poultry droppings, wheat straw and green maize in the ratio of 25:15:60 (Kamra and Srivastava, 1992). It has proved to be an excellent ingredient for cattle feeding, and the process significantly destroys harmful micro-organisms possibly present in poultry litter. It was reported that on fermentating cattle waste with straw, green maize and molasses, the pH reduced to 4.3 in 15 days of fermentation and coliform bacteria were killed and the number of clostridia decreased considerably (Kamra *et. al.,* 1984). Silage made from poultry litter, chopped root crops and banana by-products provides a balanced diet for dairy cows.

Techniques of Silage Making from By-products

The technique of silage making from by-products was extensively described by Kayouli and Lee (1998). Incorporation rate of the different ingredients to be ensiled depends on the amount of by-product(s) available; and the animal types to be fed. Thus, a high-quality silage, containing increased proportions of energy-rich ingredients such as spent grain or crop roots, should be prepared for high-yielding dairy cows, while high proportions of cassava leaves and banana pseudostems can be used when there is seasonal feed shortage and therefore when silage would compose the bulk diet during off-season feeding (Haji panayiotou, 1993, 1994).

The silage can be stored in stacked layers, packed in succession on a soil surface that has been covered beforehand with a plastic sheet or banana leaves. This heap, once finished, is then tightly covered with banana leaves or plastic sheets, pressed down by some heavy objects, which are placed on top. Packed silage in plastic bags that are tightly closed is also an effective storage method. This storage method is easy to handle and has the potential to produce high quality silage with less waste in a well-sealed bag. However, it is not recommended for coarse materials, such as banana trunk and cassava leaves, which can puncture the bag and render the contents useless. After approximately 6 weeks, the farmer can open the silo and start to feed silage to animals. Silage can be suitably preserved for as long as air is kept away from the ensiled material; it is therefore possible to store airtight silage for more than 6 months. Once the silo is open, care must be taken to cover again the ensiled material after each opening that is made to feed the animals.

The ensiling of by-products is a simple and appropriate method of conservation. It is the most effective way to improve animal feed resources through the rational use of locally available agricultural and industrial-by products likely to be available to small-scale farmers at village level.

References

Al-Rokayan, S.A.; Naseer, Z.; Chaudry, S.M. 1998. Nutritional quality and digestibility of sorghum-broiler litter silages. *Animal Feed Sci. and Technol.* 75: 65-73.

Ashbell, G. 1992. Conservation of citrus peels by ensiling for ruminant feed.In: Anais do Simpósio Utilização de Sub-produtos AgroindustriaiseResíduos de Colheita na Alimentação de Ruminantes (A. G. da Silva, etal. eds.) São Carlos, SP. p. 189-190.

Bouqué, C.V. and Fiems, L.O. 1988. Vegetable by-products of agro-industrial origin. *Livestock Prod. Sci.* 19: 97 - 135.

Chedly, K and Lee, S. 1999. Silage from by-products for smallholders. In: L.'t Mannetje (Ed.) Proc. FAO e-Conf. on Tropical Silage. 1 Sept. - 15 Dec. 1999. http://www.fao.org/AG/Agp/agpc/gp/SILAGE/PDF/Paper6.pdf

Hadjipanayiotou, M. 1987. Experience in the Near East with the utilization of non-conventional feedstuffs and the various systems of feeding. Animal Production and Health Paper no. 54; FAO, Rome, pp 185 - 210.

Hadjipanayiotou, M. 1993. Voluntary intake of citrus pulp-poultry litter silage by growing female Friesian calves, chios lambs and Damascus kids. *Technical Bull*etin. 155. Agricultural Research Institute, Nicosia.

Hadjipanayiotou, M. 1994. Voluntary intake and performance of ruminant animals offered poultry litter olive cake silage. *Livestock Research for Rural Development.* 6:10.

Kamra, D.N., Jakhmola, R.C. and Singh, R. 1984. Ensiling of cattle waste and wheat straw with or without green maize. *Agric. Waste.* 9: 309-313.

Kamra, D.N and Srivastava, S.K. 1992. Effect of premix ingredients and its inoculation with lactic acid producing bacteria on ensiling of poultry droppings. *J. Applied. Microbiol.* 73: 371-374.

Kayouli C. 1989. A strategy for animal nutrition in the dry subtropics. International Symposium on the Constraints and possibilities of ruminant Production in the Sub-Tropics (MOA of Egypt, ESAP, EAAP, FAO, ICAMS, WAAP), Cairo, Egypt, November 1988. Pudoc Wageningen, pp 201 - 206.

Kayouli C., Demeyer, D. and Accacha, M. 1993. Evaluation of poultry litter and olive cakes as an alternative feed for ruminant production in Tunisia. Proceedings of the International Conference on Increasing Livestock Production through

Utilisation of Local Resources. October 18-22, 1993 Beijing, China. pp. 420 - 440

Kayouli, C. and Lee, S. 1998. Supplementary feeding for dairy smallholders in Pacific Island Countries: Fiji, Samoa, Vanuatu, Cook Islands, Solomon Islands and Tonga. In: Manual of Smallholder Milk Production in the South Pacific, (Eds. Stephen Lee, Roderick Kennard and Chedly Kayouli), FAO Sub-Regional Office for the Pacific, Apia, Samoa, pp.: 67 - 101.

Lien, L.V., Sansoucy, R. and Thien, N. 1994. Preserving shrimp heads and animal blood with molasses and feeding them with a supplement for pigs. In: Proceedings of National Seminar-Workshop "Sustainable Livestock Production On Local Feed Resources" (Eds. T.R. Preston, B. Ogle, Le Viet Ly and Lu Trong Hieu), Ho Chi Minh City, November 22- 27, 1993. pp. 59 - 62.

Mc Donald, P. 1983. Hannah Research Institute Annual Report. pp 59 - 67. Thomas, C. 1985: In: *Recent Advances in Animal Nutrition*. (Eds. W, Haresign and D.J.A Cole). London: Butterworths, pp 223 - 256.

Pitt, R.E. 1991. Hay preservation and hay additive products. In: Field Guide for Hay and Silage Management in North America, p 127. K. K. Bolsen, (ed.) Natl. Feed Ingredients Assoc. West Des Moines, Iowa.

Satter, L.D., R.E. Muck, J.A. Woodford, B.A. Jones, and C.M. Wacek. 1988. Inoculant research: what has it shown us. In: Proc. of the Wisconsin Forage Council's 12th forage production and use symposium. WI Forage Council, Wisconsin Dells, WI. pp. 108-119.

CHAPTER 11

Losses During Silage Making

Objective of the conservation of forage is to preserve as much as of the crop nutrient as possible. However, during ensilage loss of nutrients occurs. There are five sources of loss. Theses are:

(a) Field losses

(b) Oxidation or respiration losses

(c) Fermentation losses

(d) Effluent losses

(f) Aerobic deterioration

A. Field losses

Filed loss is a result of respiration activity of the harvested crop and physical factors such as cutting, shattering, leaching by rain, and the efficiency of pick-up by the forage harvester. The main nutrients affected are the water-soluble carbohydrate, and protein which are hydrolyzed to amino acids. Prolonged period of wilting can result in substantial dry matter losses. With crops harvested and ensiled the same day, nutrient losses are negligible and even over 24-hours wilting period, losses of dry matter of 1-2% occur. Dry matter losses as high as 6% after five days, and 10% after eight days of wilting in the filed have been reported.

B. Oxidation losses

Oxidation losses result form oxidation of sugar by the plant and microbial enzymes leading to the formation of carbon dioxide and water. As the forage crop is cut and ensiled, the aerobic fermentation continues till the dry matter is increased to a level at which respiration enzymes are inactivated or sufficient acids are produced to stop the respiration or the complete utilsation of oxygen produced leading to anaerobic condition.

Sugar + oxygen $\rightarrow$ carbon dioxide + water + heat (3.8 Mcal/kg sugar)

Respiration is also responsible for protein breakdown. As plant cells die after cutting, proteolytic enzymes break down large proteins into smaller soluble compounds including: peptides, amino acids and ammonia. In addition, enterobacteria have proteolytic enzymes that remain active even after the pH has dropped to 5.0. Thus most of the protein degradation that takes place in a silo occurs within the first 24 to 72 hours. By the time pH is about 4.0, proteolytic enzymes have lost 65 to 85% of their activity. Consequently, a rapid drop in pH is desirable to reduce the amount of protein breakdown in a silo. Nevertheless, recent research indicates that as much as half of the total nitrogen in alfalfa silage may be in the form of non-protein nitrogen. The extensive breakdown of alfalfa protein in a silo may be a factor in limiting milk yield in high producing cows (Broderick, 1995). Some scientists have proposed the use of ammonia content as one indicator of adequate silage fermentation. For example, grass silage containing less than 5% of total nitrogen in the form of ammonia might be classified as excellent. In contrast, a silage that undergoes clostridial fermentation might contain more than 35% nitrogen in the form of ammonia and thus would be classified as poor (Vanbelle, 1985). Although limiting protein solubilization and ammonia production in a silo is desirable, one should realize that this process is relatively similar to what happens in the rumen of a cow after ingesting silage. Not all soluble protein and ammonia produced in the silo are necessarily lost. Ammonia and amino acids are needed for microbial synthesis in the rumen. Another nitrogen transformation that may occur during hay or silage- making results from the browning reactions that take place when excess heat is produced. Excess heat cause chemical reactions that combine amino acids with plant sugars (usually derived from hemicellulose) to form a compound resembling lignin. This reaction results in increased levels of acid detergent fiber

(ADF) and acid detergent insoluble protein. Excess heat is an indication of aerobic degradation due to the presence of air (oxygen) in a silo.

Thus respiration results in a loss of dry matter and available energy. In addition, the heat released by respiration raises the temperature of the forage. Temperatures higher than 26-32°C may cause significant loss of nutrients. Research indicates that the rise in temperature is not as high in well packed silos as it is in poorly packed silos (Pitt, 1990). Normally, respiration continues for one to two days, but takes place only as long as oxygen is present in the silage. The rapid expulsion of oxygen is desirable because it decreases both the length of the respiration phase and the associated nutrient losses. As long as aerobic condition prevails in silo, and the pH of the forage has not been reduced, respiration by both plant cells and the aerobic microflora will continue. Therefore, rapid filling of silo can reduce oxidation losses. Oxidation losses may vary between 5-10% depending upon the condition and extent of wilting the green forages. Following mangemental conditions can curtail aerobic deterioration:

- Chop the forage to a length of 1.0 cm for green forages and 0.6 cm for maize silage. Short particle length enhances sugar availability for fermentation and facilitates silo compaction. However, at least 20% of the particles should exceed 2.5 cm in length to insure enough effective fiber in the silage.
- Filling a silo should be a continuous process with delays no longer than overnight. The last load of the day should always be packed particularly well to reduce oxygen penetration overnight.
- Pack the silage as much as possible to expel oxygen and favor the growth of lactic acid bacteria.

C. Fermentation losses or ensiling losses

Fermentation loss is the result of the activities of anaerobic or facultative anaerobic microorganisms under aerobic conditions. The losses of nutrients occurring in the silo are mainly due to the activity of microorganisms. The fermentation losses mainly depend upon the ratio of glucose to fructose and the ratio of homo and heterofermentative lactic acid bacteria active in the ensiling process. In well preserved silage in which dominant microorganism is lactic acid bacteria, the DM recovery would be 100%. If one consider heterolactic fermentation of glucose, there is a 24% loss of DM through evolution of carbon dioxide. In case of clostridial and entrobacterial fermentation, losses occur due to the evolution of the gaseous carbon dioxide, hydrogen and ammonia.

Clostridia (butyric acid bacteria) fermentation

Clostridia grow in the absence of oxygen (anaerobic) and are normally found in soil and manure. The fact that clostridia live in the absence of oxygen and resist pH as low as 4.2 allows them to compete with lactic acid bacteria even after the pH drops below 5.0. Essentially, clostridia dominates the fermentation when lactic acid bacteria do not produce enough lactic acid to drop pH to a stabilization value fast enough. When silage is contaminated with dirt (e.g., from a soiled tractor wheel) or manure, the risk of clostridial fermentation increases. Clostridia tend to growth faster at a temperature of about 35°C. Thus this type of undesirable fermentation happens when extensive respiration and enterobacterial fermentation occur and silage temperatures rise in the early phases of the fermentation process. Some species of clostridia ferment sugars and change the lactic acid produced by lactic acid bacteria into butyric acid, carbon dioxide and hydrogen (H_2) as follows:

Sugar → butyric acid + carbon dioxide + hydrogen

Lactic acid → butyric acid + carbon dioxide + hydrogen

Production of carbon dioxide and hydrogen gas indicates a loss of digestible energy. The breakdown of lactic acid into butyric acid, which is a weaker acid, means that the pH of silage going through clostridial fermentation will tend to rise. Butyric acid has a strong, repulsive smell. Trace amounts of butyric acid suffice to decrease voluntary intake by cows. Also, some species of clostridia ferment amino acids, leading to the formation of toxic substances such as cadaverine and putrescine. Silage spoiled by clostridia is easily recognizable due to its strong odor, a pH above 5.0, ammonia nitrogen greater than 10% of total nitrogen, and more butyric acid than lactic acid.

Fortunately, many clostridia are more sensitive to high acidity and high osmotic pressure (high levels of dry matter content) than lactic acid bacteria. Thus clostridial fermentation can be avoided by:

- Ensiling at more than 30% dry matter content;
- Packing the silo as densely as possible to reduce increases in temperature due to respiration;

- Avoiding soil contamination;
- Ensiling forages with the highest possible sugar content;
- Using proper ensiling techniques that favor maximum fermentation to the lowest possible pH as soon as possible.

D. Effluent losses

Effluent contains sugar, soluble nitrogenous compounds, minerals and fermentable acids, which are of high nutritional value. In most silos, free drainage occurs and the liquid or effluent carries with it soluble nutrients. These effluents have generally high concentration of protein, lactic acid, soluble carbohydrates and non-nitrogenous compounds. Steen (1986) reported that effluent contained 25% crude protein, 25.7% lactic acid, 31.7% total organic acids and 19.9% ash in green grass. The quantity of effluent produced depends largely on the moisture content of the forage at ensiling. Losses of dry matter in effluents occur when silage dry matter content is less than 25%. Dry matter losses decrease from 7.2 to 1.6 and 0.4% from grass silage ensiled at 15, 20 and 25% dry matter respectively (Harrison and Fransen, 1991).Crops ensiled with a dry matter content of 15% may result in dry matter effluent loss of as high as 10%, whereas crops wilted to about 30% dry matter produce little effluent loss, if any. Miller and Clifton (1965) observed negligible effluent production at a dry matter of 32.7% of ensiled green forage and the dry matter had the following correlation with silage effluents

$D=17.614- 0.534X$

where,

D=% dry matter in the effluent

X= % dry matter in the forage crop

Effluent loss also increases if the silo is left uncovered so that rain enters.

E. Aerobic deterioration

Aerobic deterioration occurs when the surface of silage is exposed to air. Such losses may consist of surface loss and feed- out loss. Surface loss occurs at the surface and sides of silo and is more characteristics of bunker silo than tower silo. Feed-out losses occur as soon silo is opened for feeding of silage to livestock. This may occur (a) on the top surface of a silo that is not sealed with a plastic sheet; (b) locally, around a whole in the plastic covering, or (c) when less than 10-15 cm of silage is removed daily from the vertical front of a well preserved silo. Losses caused by not protecting a silo from oxygen at ensiling time are due to continued respiration and undesirable bacterial fermentation. However, aerobic deterioration can also result from the development of other microorganisms including molds and yeast. Molds and yeast (and some aerobic bacteria) resist pH as low as 2.0, but remain dormant in stable silage at a pH of around 4.0-4.5. Molds require sugar and oxygen to grow and can resume their development rapidly once oxygen is present. Yeast, on the other hand, may grow with or without oxygen and can produce alcohol in silage that is rich in sugars, such as maize. More than 60 species of molds have been isolated from silage, but the risk of aerobic deterioration depends upon the type of forage ensiled. Paradoxically, the risk of development of these undesirable micro-organisms increases with the quality of silage preservation. The residual sugar in a well preserved silage is the ideal energy source for molds and yeast. In addition, well preserved silage has little or no butyric acid, which is a strong inhibitor of mold and yeast growth. Thus, after a silo has been opened, well preserved silage has a greater chance of mold and yeast deterioration than poorly preserved silage. On aerobic deterioration the levels of acetic acid, lactic acid and ethanol decreases considerably while the pH and number of total yeasts and lactate utilizing yeasts increase significantly in comparison to that in silage taken from recently opened silo (Selmer-Oslen *et al.,* 1993).

Following mangemental conditions can curtail aerobic deterioration:

- Seal the open surface of the silo with a plastic sheet to prevent oxygen from reentering the silo; poor covering (or no covering at all) causes large losses of dry matter, undesirable fermentation and unpalatable silage. Plastic sheet should have a low level of permeation by air and be carefully placed and anchored to seal the silo completely.
- Leave the silo close for at least two weeks for fermentation to reach its stabilization phase.
- Plan to build silos such that, upon unloading, 5 to 10 cm of silage can be removed daily (so that the rate of dry matter removal is greater than the rate of oxygen penetration in the silo's front).

- Use equipment that leaves a smooth surface to minimize exposure to oxygen and risk of secondary fermentation.
- When unloading a silo, remove only the amount needed for one meal (one day) and avoid spoilage of silage in feed bunks (mangers) by cleaning them every day before offering freshly unloaded silage to cows.

Table 1: Average losses and quality changes in covered and uncovered bunker silos (Oelberg *et al.* 1983)

	Temp. (°C)	DM Losses (%)	pH	Lactic acid (% DM)	CP (% DM)	ADF (% DM)	SP (% CP)	ADIP (% CP)
Covered	36.7	4	4.9	3.2	21	39	20	28
Uncovered	53.9	32	6.8	1.7	22	47	16	37

DM = dry matter; CP = crude protein; ADF = acid detergent fiber; SP = soluble protein; ADIP = acid detergent insoluble protein.

Losses during silage making

Ensiling losses Source	Classification	% Net energy lost	Cause
Respiration loss	Unavoidable	1-2	Plant enzymes, aerobic bacteria
Wilting loss	Unavoidable	2-5	Continued respiration
Fermentation (butyric acid) loss	Unavoidable	4	Heterofermentative bacteria
Secondary Fermentation loss	Avoidable	0-5	Clostridia
Effluent loss	Avoidable	5-7	Low DM at harvest
Surface wastage/ Aerobic spoilage	Avoidable	0-10	Aerobic microbes

References

Broderick , G.A. 1995. Performance of lactating dairy cows fed either alfalfa silage or alfalfa hay as the sole forage. *J. Dairy Sci.* 78:320-329.

Harrison, J.H. and Fransen, S. 1991. Silage management in North America. In Field Guide for Hay and Silage Management in North America, pg. 33. K. K. Bolsen, (ed). National Feed Ingredients Assoc. West Des Moines, Iowa.

Miller, W.J and Clifton C. M. 1965. Relation of dry matter content in ensiled material and other factors to nutrient losses by seepage. *J. Dairy Sci.* 48: 917-923.

Oelberg, T.J., Clark., A.K. McGuffey, R.K and Schingoethe, D.J. 1983. Evaluation of covering, dry matter and preservative at ensiling of alfalfa in bunker silos. *J. Dairy Sci.* 66:1057-106s8.

Pitt, R.E. 1991. Hay preservation and hay additive products. In Field Guide for Hay and Silage Management in North America. pp. 127.

Steen, R.W.J. 1986. An evaluation of effluent from grass silages as feed for beef cattle offered silage based diets. *Grass and Forage Science.* 41:39-46.

Selmer-olsen, I., Henderson, A. R., Robertson, S and Mcginn, R. 1993. Cell wall degrading enzymes for silage. 2. Aerobic stability of enzyme-treated laboratory silages. *Grass and Forage Science*. 48:55-63.

Vanbelle, M. and Bertin, G. 1985. L'ensilage, aspect biologiques nouveaux. Université deLouvain, Faculté des Sciences Agronomiques, Place croix du Sud, 3 Louvain La Neuve, 1348, Belgium.

CHAPTER 12

Feeding Silage to Livestock

Feeding silage is also an important part of the process. Silage can be fed to all kinds of animals. Silage is suitable for feeding 6-7 weeks after ensiling. Purposes of silage feeding are drought feeding, production increases, an aid to pasture or crop management, utilization of excess growth, balancing nutrients in the diet, and the storage of wet feed products. Silage is most likely to be profitable when used to increase productivity or balance nutrient content of the diet.

Feeding value of silage is defined as the product of the nutritive value and potential voluntary intake of the forage (Wheeler and Corbett, 1989). Before feeding, silage sample should be analyzed to determine the quality of the silage. This will allow calculating the level of supplementation needed according to the type of animal and it nutritional requirements. It must be supplemented with protein, minerals, and sometimes energy to meet the animal's nutrient requirements. High quality silages are capable of supporting the high levels of animal production demanded and, are often lower in cost than grains and hay, enable higher productivity from land, and maintain a more stable rumen environment. Cowan *et al.* (1991) evaluated the effects of incorporating maize silage into a grazing system on the productivity of a dairy farm. They reported productivity increase by 21,000 to 150,000 L milk/ farm/year above the previous system based solely on grazed pasture. In a separate survey, Kerr and Chaseling (1992), observed an increase in milk yield of 0.73 L for each kg hay or silage dry matter used in the feeding program. Much of the silage is made from crops during summer, and fed in the autumn and spring periods when pasture supply is normally low. This results in a relatively stable pattern of production throughout the year.

Among silages, maize silage is most popular. Although maize silage is occasionally used as the only forage for dairy cattle, it is usually fed with complimentary forage such as alfalfa which is higher in crude protein but lower in energy. Maize silage feeding strategies vary depending on animal age, level of production, and physiological status as well as the other forages being fed, if any. Because of its high grain content, feeding strategy for maize silage fed to high producing cows differs from most other forages. Maize silage is high in starch (35%) which is usually highly digested in the rumen. Excessive starch can lead to digestive upsets and low energy intakes due to excess accumulation of fermentation acids in the rumen. Supplementing maize silage with a starch source such as high-moisture maize that is highly digestible in the rumen is likely to cause digestive upset, therefore, to optimally supplement forages, one must consider the starch content and ruminal fermentability of grains as well as forages. To increase energy intake, maize silage should be supplemented with a grain source that partially bypasses ruminal fermentation yet has high whole-tract digestibility such as dried ground maize. When maize silage is fed in combination with alfalfa, less bypass starch is needed as the fraction of alfalfa is increased.

Maize silage is excellent dry cow forage as it is palatable and can be fed mixed with lower energy forages. However, the amount of maize silage fed to dry cows and heifers must be restricted due to its high energy content. Feeding unrestricted quantities of maize silage throughout the dry period will result in fat cows which tend to have lower dry matter intakes and higher incidence of metabolic problems such as ketosis and fatty liver following calving and also lower milk production due to fatty infiltration of the udder. A solution to restricted feeding of maize silage to heifers is to feed a combination of maize silage and a higher protein, lower energy forage such as legume silage. Maize silage contains low calcium compared to other forages. This is an advantage because high calcium intake in the dry period contributes to milk fever following calving. Silage quality factors that are important to consider when balancing rations are energy content, NDF content, NDF digestibility, length of cut, starch content, and starch digestibility.

The amount of silage offered depends on the animal and its age, as well as on the type and quality of silage. Taking cattle as an example, the recommended amount to be fed daily is:

Grass silage	4 kg per 100 kg liveweight,
Legume silage	3 kg per 100 kg liveweight,
Maize stover silage	4 kg per 100 kg liveweight,
Starch-rich forage (whole-maize silage)	5 kg per 100 kg liveweight,

Therefore, a cow with live weight of 500 kg may be daily fed 20 kg maize stover silage. Some animals may not like silage when offered for the first time. In these cases, some adaptive measures should be taken. For example, silage can be placed in the bottom of the trough, and covered with concentrates. When animals are adapted to silage, the offer can be increased.

Feed out rate management of silage bags

Nutritive losses occur due to exposure of silage surface. Data has shown that approximately 50% of the dry matter losses occurs from secondary anaerobic while silage is exposed to oxygen when in storage or in the bunk. This usually can lead to production of high populations of yeast and mold. The basic principle for silage use is to minimize the area exposed to air. Silage should be taken from one side of the trench, and the exposed surface should be immediately covered to minimize re-entry of air and avoid secondary fermentations. The removal rate (inches removed from the silage face per day) influences loss during the feed out period. Maintaining adequate removal rates are especially critical with hay crop silages, high moisture maize and drier silages. When designing storage, a good recommendation is to plan to remove at least twice the minimum recommended removal rate. Muck and Holmes (2006) recommended a minimum silo bag removal rate of 12 inches per day. Material taken out from the silo should be fed as soon as possible, since it rapidly deteriorates upon contact with air.Only the amount of silage that will be fed in a short period should be removed or uncovered at one time. At no time should more than three days worth of silage be exposed. With bags, it is always important to close the bag off after each feeding. Open bags cause billowing plastic that pumps air over the silage, increasing losses. The distance to which oxygen can penetrate is usually determined by how well the silage was packed. Therefore, the silage face should remain tight and smooth (Clark, 2002). Silage left in the trough must be cleaned away in time to prevent putrefaction.

Silage Fermentation Products and Voluntary Feed Intake

It is generally recognized that voluntary intake of silage is less than that of the same forage that has not undergone fermentation. The intake of silage by dairy cows declines as the concentrations of silage ammonia and butyric acid increase.

Following factors affect voluntary feed intake of silage:

Fermentation acids

Butyric acid was first implicated as being responsible for reducing silage intake (Harris and Raymond, 1963). There is moderate correlation between fermentation acids and voluntary intake (Wright *et al.* 2000). Cushnahan *et al.* (1995) reported that silage intake was 27% less than intake of the same forage fed without ensiling. Buchanan-Smith (1990) assessed the effect of adding individual or combinations of VFA and ammonia on silage palatability by measuring intake for 30 min after a meal. They reported fermentation end-products were unimportant for voluntary feed intake. While acetic acid did reduce intake when supplemented alone, other VFA, lactic acid and ammonia had no consistent effects. A creative approach to assessing the importance of fermentation acids on silage DM intake was adopted by Dawson and Mayne (1998). These authors squeezed juices from two silages of diverse fermentation quality and recombined these juices with the alternate silages. No relationship was found between organic acid concentration and silage DM intake.

Silage pH

Low pH in silages is often associated with poor intakes because low pH in the rumen reduces cellulolytic activity and depresses intake. However, there is no relationship between silage pH and rumen pH (Rooke, 1995). Silage is neutralized by saliva upon consumption. Rooke (1995) and Offer (1997) have postulated that the free acid content of silage, as distinct from pH, is related to intake. However, Rooke (1995) also suggested that lactic acid may had a direct effect on palatibility, since sour taste is associated with reduced palatibility.

Ammonia N

Ammonia nitrogen in silage has long been associated with reduced silage intake (Cushnahan and Gordon, 1995; Patterson *et al.* 1996). In part this has arisen because it can be readily measured, and may act as a simple index of silage fermentation quality. Ammonia N in silage is predominantly a product of clostridial fermentation of amino acids. However, many of the other products of amino acid breakdown can reduce intake (Buchanan-Smith, 1990).

Protein solubility

Ruminal ammonia levels, on the other hand, may have an impact on silage DM intake. After feeding silages, ruminal ammonia concentration can increase to 80 mg/dl (Charmley and Veira, 1990a). These very high levels are not related to the level of ammonia in silage, but to the amount and solubility of CP in silage. Thus, silages with a high CP content and high solubility, such as alfalfa, can result in high rumen ammonia concentration. Under certain feeding situations, these conditions could lead to mild ammonia toxicosis which may reduce feed intake (Choung *et al.*, 1990). Choung *et al.* (1990) suggested that there should be a strong relationship between non-protein N in silage and Silage DM intake. Initially, increasing solubility leads to increases in DM intake and BW gain. However, as solubility increases above 475 g/kg-1 total N then intake and gains decline markedly. Charmley and Veira (1990a and b) used high temperature-short term heat treatment to denature plant protease enzymes in alfalfa. This treatment dramatically reduced the solubility of CP in the silage and increased intake. Similar responses in DM intake by reducing protein solubility have been achieved when rapid acidification (Charmley *et al.*, 1995) or wilting (Muck, 1987) have been used to inhibit proteolytic activity. These data suggested that there is a relationship between silage protein solubility and voluntary intake and BW gain.

Measures to Improve Silage Intake and Utilization

Restricting fermentation through wilting

There are two means of restricting fermentation; wilting and the use of acid-type additives. Wilting indiscriminately reduces microbial activity in silage, thus restricting the extent of fermentation. However, it also tends to favour lactic acid bacteria over other types, thus reducing the risk of less useful microbial pathways dominating the ensilage process (McDonald *et al.* 1991).

It was also suggested that intake of silage was increased by wilting (Marsh, 1979; Zimmer and Wilkins, 1984). It is known that wilting markedly reduces proteolysis by plant enzymes in the silo

(Muck, 1987). The faster the rate of drying, the more effective wilting is at reducing proteolysis (Anderson, 1983). Animal studies have linked this reduction in protein solubility to increased amino acid flow to the intestines (Charmley and Veira, 1990 a,b), and milk yield in lactating cows (Broderick *et al.*, 1993). Thus, the benefits of wilting on animal performance may be related primarily to an improvement in the utilization of N by the ruminant. It would appear that methods to increase drying rate will have a profound effect on animal performance. Such methods include crop spreading, to expose the maximum surface area to solar radiation (Wilkinson *et al.* 1999), and crop conditioning to reduce plant resistance to moisture loss (Savoie *et al.*, 1993; Frost and Binnie, 1999). Wilkinson *et al.*, (1999) increased drying rate by spreading swaths during wilting. Such swaths reached a higher DM concentration by ensiling. A response in intake by dairy cows was observed, but it is difficult to conclude if this was due to drying rate or the DM content at which forage was ensiled.

Restriction of fermentation through acidification

Acid-type additives have a similar effect on proteolysis to wilting by inhibiting plant protease enzymes through rapid reduction in pH (Charmley *et al.*, 1995; Keady and Murphy 1997). Charmley *et al.* (1994) used three levels of acid salts on wilted alfalfa and proved that proteolysis was progressively restricted as the level of additive was increased. This translated to improved growth of steers fed the same quantity of silage. Nagel and Broderick (1992) increased insoluble protein in alfalfa silage with formic acid and this increased milk production by 12%. Added acids are more effective than natural fermentation because acidification occurs within minutes of adding the additive. When relying on natural fermentation, acidification can take days or weeks.

Enhancing silage fermentation by inoculation and enzymes

Inoculants and enzymes have become popular as a means of improving silage nutritive value as they increase the extent of fermentation in the silo. Cellulase, hemicellulase and amylase enzymes have been widely tested as silage additives. These compounds have the potential to convert structural carbohydrate to soluble sugars which can be fermented by silage bacteria. Their use increases the level of fermentable substrate in silage, thus promoting extensive fermentation (Jacobs and McAllan, 1991; Stokes, 1992).

It was concluded that inoculation increased animal performance from silages (Spoelstra, 1991; Muck, 1993). The improvement in performance was due to an increase in digestibility, rather than an increase in intake. Since lactic acid bacteria are non-cellulolytic. The increase in digestibility may be in response to the extensive fermentation of substrate to lactic acid in these silages. Under these conditions, acid hydrolysis of structural carbohydrate to soluble sugars will occur (Dewar *et al.* 1963). The reduction in fibre concentration has been speculated to be responsible for increases in digestibility (Heron and Owen, 1991). On the other hand, Gill and Romney (1994) and Teller *et al.* (1993) have demonstrated that highly fermented silages are eaten more slowly than less highly fermented silages or hay. If these changes in eating behaviour influence ruminal retention time, then this could also explain why digestibility is increased by inoculant use.

Problems Associated with Silage Feeding

Off smell in milk

Sometimes the milk from cows fed on silage has a taint. Because this smell is only transmitted through the air, the following points should be observed to prevent it:

1. *Handling:* Silage should never remain in the cowshed, and it should be offered in amounts exactly as required.
2. *Feeding method:* Silage should not be given before milking. The effect on milk taste is more marked when the silage is fed 2 hours before milking, and least when given 6 hours after.
3. *Hygiene and cleanliness:* Both floor and operators should be clean and the shed should be well ventilated.
4. *Milk and milking:* Equipment should be kept clean, and milk should be cooled as soon as possible.

Nutritional deficiency

It is sometimes believed that ensiled forage has an adverse effect on the general health of the animal, especially on the skeleton of young animals. It must be pointed out that silage is a good feed, but it is certainly not a "complete feed." Animals fed on silage should be properly supplemented with the nutrients that are insufficient for the animals' requirements. For example, maize silage contains insufficient calcium and phosphorus, and should be supplemented with calcium and phosphorus. It must also be emphasized that the nutritional value of silage depends on the quality of the original forage, harvest time, conservation method, etc. Nutritional values of different silages therefore vary substantially.

Mycotoxicosis

Mycotoxins are toxic substances produced by molds growing on grain or feed. While the vast majority of molds that can grow on silage are harmless, a few species produce mycotoxins. The primary mycotoxins found in maize silage are aflatoxin, deoxynivalenol (DON or vomitoxin), zearalenone, T-2 toxin, fumonisin, and ochratoxin. Mycotoxins can cause illness or death to livestock and humans. More commonly, contaminated feed reduces growth rate, lowers feed conversion and reproductive rate, impairs resistance to infectious diseases, and reduces vaccination efficacy.

Mycotoxin-contaminated feed poses a greater health risk to young or pregnant animals and animals that are stressed. The maximum dietary level for aflatoxin has been set by the Food and Drug Administration (FDA) for cattle to be no more than 200 parts per billion (ppb) for breeding cattle, 300 ppb for finishing cattle, 20 ppb in feeds for lactating dairy cattle. Never use contaminated feeds that exceed these levels.

Prevention is the most cost-effective and safest way to avoid problems associated with mycotoxins. Since mold contamination begins with the growing plant, sound agronomic practices which minimize the incidence of insect and disease levels will reduce the potential contamination in the silage. Often environmental factors out of the producer's control such as wind and hail damage, drought, or early frost, will dictate the use of the crop for silage. Under these conditions, the mold level on plants will be elevated and this increases the potential risk for mycotoxin production. The key to the prevention of mold growth during the ensiling process is the elimination of air as quickly as possible, and managing the silo to minimize air infiltration into the ensiled mass during storage and feedout. Employing sound silage management practices including rapid filling, harvesting at the correct moisture level, adequate compaction, covering exposed surfaces, and rapid feedout will go a long way in minimizing mold growth and potential mycotoxin contamination in silages. The use of some silage additives may be beneficial in reducing the risk of potential mycotoxin production if they are shown to reduce mold growth. Ammonia and/or propionic acid appear effective in this regard. Certain bacterial inoculants may also be beneficial since they have been shown to increase the rate of pH decline in the silage and extend aerobic stability of the silage on feedout.

At present there is no practical way to detoxify mycotoxin-contaminated silages. If unacceptably high levels of mycotoxins occur, dilute or eliminate the silage from the diet. The addition of absorbent materials, such as clays (bentonites) or anticaking agents (hydrated sodium calcium aluminosilicate), to contaminated rations has helped in some cases and may warrant consideration. Avoid feeding moldy clumps of silage or visibly moldy silage, particularly to pregnant animals or young stock.

References

Anderson, R. 1983. The effect of extended moist wilting and formic acid additive on the conservation as silage of two grasses differing in total nitrogen content. *J. Sci. Food Agric.* 34:808–818.

Broderick, G.A., Craig, W.M. and Ricker, D.B. 1993. Urea versus true protein as a supplement for lactating dairy cows fed grain plus mixtures of alfalfa and corn silages. *J. Dairy Sci.* 76: 2266–2274.

Buchanan Smith, J.G. 1990. An investigation into palatability as a factor responsible for reduced intake of silage by sheep. *Anim. Prod.* 50: 253–260.

Charmley, E. and Veira, D.M. 1990a. Inhibition of proteolysis at harvest using heat in alfalfa silages - effects on silage composition and digestion by sheep. *J. Anim. Sci.* 68: 758–766.

Charmley, E. and Veira, D.M. 1990b. Inhibition of proteolysis in alfalfa silages using heat at harvest - effects on digestion in the rumen, voluntary intake and animal performance. *J. Anim. Sci.* 68: 2042–2051.

Charmley, E., McQueen, R.E. and Veira, D.M. 1994. Influence of carboxylic salts (Maxgrass) on silage conservation, and voluntary intake and growth of steers given lucerne silage. *Anim. Prod.* 58: 221–229.

Charmley, E., Veira, D.M., Berthiaume, R. and McQueen, R.E. 1995. Effect of a mixture of carboxylic salts on silage conservation, and voluntary intake and growth of cattle given grass silages. *Can. J. Anim. Sci.* 75: 397–404.

Choung, J.J., Chamberlain, D.G, Thomas, P.C. and Bradbury, I. 1990. The effects of intraruminal infusions of urea on the voluntary intake and milk production of cows receiving grass silage diets. *J. Dairy Res.* 57: 455–464.

Clark, J. 2002. Forage Feed out Losses for Various Storage Systems. University of Wisconsin Extension, Team Forage, Focus on Forage Factsheet.. http://www.uwex.edu/ces/crops/uwforage/storage.htm

Cowan, R.T., Kerr, D.V. and Davison, T.M. 1991. Maize silage for dairy systems in northern Australia. In: Maize in Australia - Food, Forage and Grain. (Ed. J Moran). Incitec, pp. 228-235.

Cushnahan, A. and Gordon, F.J. 1995. The effects of grass preservation on intake, apparent digestibility and rumen degradation characteristics. *Anim. Sci.* 60: 429–438.

Dawson, L.E.R. and Mayne, C.S. 1998. The effect of silage fermentation characteristics on dry-matter intake of steers. *Anim. Sci.* 66:105-113.

Dewar, W.A., McDonald, P. and Whittenbury, R. 1963. The hydrolysis of grass hemicelluloses during ensilage. *J. Sci. Food Agric.* 14: 411–417.

Frost, J.P. and Binnie, R.C. 1999. The effect of mechanical treatment on the drying rates of Italian and perennial ryegrasses. *Grass For. Sci.* 54: 144-154.

Gill, M. and Romney, D. 1994. The relationship between the control of meal size and the control of daily intake in ruminants. *Livest. Prod. Sci.* 39: 13–18.

Harris, C.E. and Raymond, W.F. 1963. The effect of ensiling on crop digestibility. *J. Brit. Grassl. Soc.* 18: 204–212.

Heron, S.J.E. and Owen, T.R. 1991. Review of the effects of 'Ecosyl' silage inoculant on in vivo digestibilities and metabolizable energy of grass silages. In "Management Issues for the Grassland Farmer in the 1990's. C.S. Mayne, (ed.). British Grassland Society, Hurley, UK. pp. 230–233

Jacobs, J.L. and McAllan, A.B. 1991. Enzymes as silage additives.1. Silage quality, digestion, digestibility and performance in growing cattle. *Grass For. Sci.* 46: 63–73.

Keady, T.W.J. and Murphy, J.J. 1997. The effects of treating low dry matter herbage with a bacterial inoculant or formic acid on the intake and performance of lactating dairy cattle. *Anim. Sci.* 64: 25-36.

Marsh, R. 1979. The effects of wilting on fermentation in the silo and on the nutritive value of silage. *Grass For. Sci.* 34: 1–10.

McDonald, P, Henderson, A.R. and Heron, S.J.E. 1991. The Biochemistry of Silage. Second Edition, Chalcombe Publications, Bucks, UK.

Muck, R. and B. Holmes. 2006. Bag Silo Densities and Losses. *Transactions of the ASABE.* Vol 49(5): pp. 1277-1284.

Muck, R.E. 1987. Dry matter effects on alfalfa silage quality. 1. Nitrogen transformation. *Trans. Amer. Soc. Agric. Eng.* 30: 7–14.

Muck, R.E. 1993. The role of silage additives in making high quality silage. Pp. 106–116 *in* Silage Production from Seed to Animal, Proceedings from the National Silage Production Conference, Syracuse, NY.

Nagel, S.A. and Broderick, G.A. 1992. Effect of formic acid or formaldehyde treatment of alfalfa silage on nutrient utilization by dairy cows. *J. Dairy Sci.* 75: 140–154.

Offer, N.W. 1997. A comparison of the effects on voluntary intake by sheep of dietary addition of either silage juices or lactic acid solutions of the same neutralizing value. *Anim. Sci.* 64: 331- 337.

Patterson, D.C., Yan, T and Gordon, F.J. 1996. The effects of wilting of grass prior to ensiling on the response to bacterial inoculation. 2. Intake and performance by dairy cattle over three harvests. *Anim. Sci.* 62: 419-429.

Rankin, M. 2001. Silage Density and Dry Matter Loss in Silo Bags. University of Wisconsin Extension. http://www.uwex.edu/ces/crops/

Rooke, J.A. 1995. The effect of increasing acidity or osmolality of grass silage by the addition of free or partially neutralized lactic acid on silage intake by sheep and upon osmolality and acid-base balance. *Anim. Sci.* 61: 285-292.

Savoie, P., Binet, M., Choiniere, G., Tremblay, D., Amyot, A. and Theriault, R. 1993. Development and evaluation of a large-scale forage mat maker. *Trans. Amer. Soc. Agric. Eng.* 36: 285–291.

Spoelstra, S.F. 1991. Chemical and biological additives in forage conservation. *Landbauforschung Völkenrode* 123: 48–70.

Stokes, M.R. 1992. Effects of an enzyme mixture, an inoculant, and their interaction on silage fermentation and dairy production. *J. Dairy Sci.* 75: 764–773.

Teller, E., Vanbelle, M. and Kamatali, P. 1993. Chewing behaviour and voluntary grass silage intake by growing cattle. *Livest. Prod. Sci.* 33: 215–227.

Wheeler, J.L. and Corbett, J.L. 1989. Criteria for breeding forages of improved feeding value: results of a Delphi survey. *Grass Forage Sci.* 44: 77-83.

Wilkinson, J.M., Hill, J. and Leaver, J.D. 1999. Effect of swath treatment on water loss during field wilting and on feeding value of perennial ryegrass silage. *Grass For. Sci.* 54:227-236.

Wright, D.A., Gordon F.J., Steen, R.W.J. and Patterson, D.C. 2000. Factors influencing the response in intake of silage and animal performance after wilting of grass before ensiling: a review. *Grass For. Sci.* 55:1-13.

Zimmer, E. and Wilkins, R.J. 1984. Efficiency of silage systems: a comparison between unwilted and wilted silages. *Landbauforschung Völkenrode.* 69. pp.88.

CHAPTER 13

Usefullness of Silage

Forage preservation as silage is a key component of high input (zero-grazing) systems (Mannetje, 1999). It has allowed producers to intensify the productivity of the land and the productivity of the livestock. Large quantities of forage can be conserved in a short time, forage conservation is less weather dependent and thirdly, silage is well suited to mechanization. However, a major disadvantage associated with silage making is that the feeding value of the resultant forage is reduced relative to that of the original crop. Ensiling offers many advantages over haymaking.

The Advantages of silage are as follows:

1. **As a drought reserve**: Fodder conservation has the main objective of ensuring feed availability during periods of scarce feed supply (Najib *et al.,* 1993). Silage can be efficiently used for strategic off-season feeding. Silage is made from pasture or crop in times of plenty and stored for a period of 1 to 20 years. The silage can be fed to animals in times of extreme feed deficiency. Silage can be efficiently used as strategic off-season feeding

2. **To increase productivity:** Silage is routinely made as a means of increasing the amount of feed available to cows. The storage period is consistently less than 1 year, and the practice is often associated with a change from pastures to crops as a form of land use. Silage is a means of increasing feed resource availability and form of insurance, especially for calving dairy cows.

3. **To aid in the management of pasture or crop**: Where the pasture or crop is removed as silage to enable benefits to be accrued from other management practices. Examples are the increased tiller density and production of temperate pastures when excess growth is removed early in the growing season, and the removal of a crop to enable the earlier planting of a subsequent crop. It can be fed to reduce pressure on pasture when required

4. **The use of excess growth**: Where the rationale is that it is a waste to allow excess growth of forage to mature and decay *in situ*, and it should be harvested for use in the future.

5. **To balance the nutrient content of the diet**: Silage can provide a major diet source, as basal ration as well as a feed supplement for grazing animals. Silage is made with the intention of feeding it to provide nutrients lacking in feeds available at that time. Examples are the use of legume silage to feed with maize silage, maize silage to feed with grazed legume pastures, or silage of relatively high fibre content to feed with pastures of low fibre content. It can be an efficient supplement to grazing cattle during the dry season. Maize silage has 30-50% higher nutritive value compared to maize grain and maize straw.

6. **To enable storage of potentially unstable material**: Ensiling process ensures the feed can be used over an extended period. An example is the ensiling of wet byproducts. This use is similar to that in the preservation of feeds through the addition of chemicals or exclusion of air from feeds such as high moisture grains.

7. **To improve palatability**: Animals prefer soft green fodder than dry grass and ensiling keeps the grass soft. When the grass is over matured with stems, it is rejected by animals. If the green grass is cut into small bits and preserved as silage, it becomes soft and is eaten by the animals. This helps to avoid wastage of grass. It reduces significantly toxic substances present in some fresh vegetables to safe level concentrations (such as cyanogenic glucosides in fresh cassava leaves) and destroys harmful micro-organisms possibly present in poultry litter or fish wastes.

8. **To save space:** Storage of dry fodder requires more space. Large amount of green grass can be stored in small silo pits of one cubic meter (Banerjee, 1998). Silage requires 10 times less storage space compared to hay.

9. **Storage of non-forage feeds**: Storage of non-forage feeds that cannot be preserved as hay, such as agro-industrial byproducts (brewers' grains, sugar beet pulp, etc.); Ensiling allows by-products (from sugar beat processing, potato waste, etc.) to be optimally used. A good quality silage can be prepared by fermentating poultry droppings, wheat straw and green maize in the ratio of 25:15:60 (Kamra and Srivastava, 1992).

10. **Less weather dependent:** Silage making is less dependent than hay-making on good weather conditions and can be extended to a great variety of forage crops (maize, sorghum, immature cereal grains, etc.) and locally available agro-industrial by products (sugar beet pulp, brewers grain, etc.). Silage is produced in both cold and cloudy weather (Woolford, 1984).

11. **Weed control**: Weeds can be ensiled along with the main forage crops. The ensiling process destroys the germinating power of weeds (Kamra and Pathak, 1997).

Drawbacks of silage making

Silage making also has some limitations or drawbacks that need to be accounted for when assessing this technology:

1. Silage making requires considerable capital and labor investments on the farm.
2. Silage also demands a fairly high level of technical expertise.
3. Silage making requires high capital investment for construction of a silo.
4. Harvesters are needed to chop the forage, tractors or other heavy equipment are used to pack the silage, storage facilities (silo structures) may be expensive, and additional equipment may be required to remove silage from a silo.
5. The management of silos is sometimes difficult on the farm because once a silo is opened, silage should be removed on a daily basis (to minimize loss of nutritive value).
6. Adjusting the number of silos and their dimension to the expected feed out rate for a given herd size is difficult. Usually, only large herds can afford to feed out of different silos of varying forage qualities for different groups of animals on the farm (heifers, dry cows, lactating cows, etc.).

7. Once silage is removed from a silo, it becomes unstable (because of exposure to oxygen) and tends to spoil within a day or two (especially in warm weather conditions with silages that are well-preserved).
8. Silage cannot be marketed easily (difficult to transport long distances compared to hay).
9. Loss of nutrients during storage in a silo is unavoidable and may be high if the silage is not prepared properly. The major losses associated with silage making include.
    i. The potential loss of highly digestible dry matter due to soluble sugars in the effluents ("juices") when silage has a low dry matter content.
    ii. The inevitable loss of protein quality (especially with certain legumes such as alfalfa) as protein peptides and amino acids are converted to soluble nitrogen (e.g., amino acids and ammonia) during the fermentation process.
    iii. The inevitable loss of energy as sugars are converted into organic acids, carbon dioxide (and other gases) and heat.

References

Banerjee, G.C. 1998. Conservation of forage crops. In: *Feeds and Principles of Animal Nutrition*.Oxford and IBH Publishing Co. Pvt. Ltd., New Delhi, India. pp. 35-51.

Mannetje, L. 1999. Introduction to the Conference on Silage Making in the Tropics. In: L.'t Mannetje (Ed.) Proc. FAO e-Conf. on Trop. Silage. FAO Plant Prod. and Protect. Paper 161. 1 Sept. - 15 Dec. 1999. Paper 1.0: 1-3.

Kamra, D.N and Pathak, N.N. 1997. Nutritional Microbiology of Farm Animals. Vikas Publishing Pvt. Ltd., New Delhi.pp. 212.

Kamra, D.N and Srivastava, S.K. 1992. Effect of premix ingredients and its inoculation with lactic acid producing bacteria on ensiling of poultry droppings. *J. Applied. Microbiol.* 73: 371-374.

Najib, M.A., Aminah, A. and Idris, A.B. 1993. Forage conservation for livestock smallholders in Malaysia. In: Strategies for suitable forage-based livestock production in Southeast Asia. Proceedings of the Third Meeting of the Regional Working Group on Grazing and Feed Resources of Southeast Asia. 31 January-6 February 1993, Khon Kaen, Thailand. p. 103-109,

Woolford, M.K.1984. *The Silage Fermentation*. Marcel Decker Inc., New York, USA. pp. 19.

CHAPTER 14

Hay Preservation and Utilization

Feeding of green and succulent fodder is of utmost importance to farm animals. Grazing fresh forage year round at the optimal stage of maturity would supply the highest quality and most palatable form of feed in any livestock operation. In most part of the world, green fodder is limited to a particular season only. Because of fluctuations in seasonal growth and plant maturity, excessive biomass in the spring, it is necessary to harvest and store forages to maximize both quality and productivity. One of the important form of conservation practiced on a wide scale is hay making. Hay making is an art in which the fodder is kept for future use in lean period. Since it contains large quantity of moisture it is to be dried properly to increase its keeping quality without reducing its nutritive value. The objective of hay making is to achieve a rapid moisture loss after cutting so that the forage can be romoved from the field with minimum losses from weathering and microbial degradation (Bareebam, 1992).

Hay refers to forage that are harvested and dried and stored as 85-90% dry matter. The forage crop is cut before it is fully ripe and dried for storage as hay. Hay is more nutritious and palatable than straw. It is leafy pliable green and free from mould, weeds and dust and has a pleasant characteristic smell and aroma. It should contain less than 15% moisture. Legume hay, non-legume hay and mixed hay are the three major types of hay used in livestock feeding. Hay can be used as an excellent feed for dairy cattle particularly when fodder is cscarce and pasture is insufficient.

Principles of hay making

The aim of hay making is to reduce the moisture level of the green crop to a level low enough so that it can be safely stored in mass without undergoing fermentation or becoming mouldy. Drying of forage inhibits the action of plant and microbial enzyme. The process of drying the green crop without significant change in aroma, flavour and nutritive quality of forage is called "curing". This involves reducing the moisture content of green forages, so that they can be stored without spoilage or further nutrient loss. Green forage with 80-85% dry matter preserves most of the nutrients.

Steps of Hay Making

1. Selection of suitable plant varieties

Selection of suitable plant varieties and their cultivation on good soil with proper care in the form of watering, fertilization etc. is the important aspect of hay making. Crops with thin stems and more leaves are better suited for haymaking as they dry faster than those with thick stem and small leaves. The seed rate should be high to ensure dense stand of crop which will result in finer stems, heavier yield and less weeds.

2. Harvest at proper stage of plant maturity

A good forage crop is only the first requisite in producing a quality hay crop. The way forage is harvested and stored determines how well the quality of the standing crop is preserved. The stage of growth of plant is the most important consideration in harvesting the forage crops. When the crop is cut at early stages, the yield is less but the product will be more nutritious. Smaller yields and frequent cuttings increase labour cost. Therefore plants are harvested for hay at an intermediate stage when neither the quality nor the quantity is at its peak. Forage needs to be cut before it is fully mature (long before it has seeded) to maximize its nutritive value. As plants mature, their lignin content (a component of fiber) increases and traps the nutrients within indigestible cell walls. Leguminous fodder crops (e.g. Cow pea, Berseem, Lucerne, etc) should be harvested at the flower initiation stage or when crown buds start to grow, while grasses (e.g. oats) and similar fodder crops should be harvested at the pre-flowering stage. At this stage, the crop has maximum nutrients and green matter. After flowering and seeding, grasses contain fewer nutrients.

In order to make the process of curing easier, the fodder should preferably be harvested when air humidity is low. Cut the fodder in the evening (but do not spread) or in the early morning. Natural physiological processes in plants cause the concentrations of soluble carbohydrates and other highly digestible nutrients to peak in the evening. Recent research suggests that hay cut at or near sundown is higher in energy than hay harvested at sunup.

3. **Curing**

Curing is the main process of hay making. In curing hay the objective is to preserve maximum nutrients and to remove moisture sufficiently to avoid moldiness and spoilage. When the forage is cut, it needs to be laid out in the sun in as a thin layer as possible, and it should be raked a few times and turned regularly to hasten drying. If there is sufficient labour, then chopping the forage after it has been cut will hasten drying. Leaves are more nutritious than the stems, and so when cutting forage, it is important that it is cut with as much leaf and as little stem as possible. However, during drying, the leaf (being more brittle) will tend to shatter. Hay should therefore be handled with care, to try and minimize the amount of leaf that is lost in this way. The dried forage is then collected and baled when the moisture content is low, ideally less than 15%. This helps storage and requires less space. Crops with thick and juicy stems can be dried after chaffing to speed up the drying process and to prevent loss of nutrients.

The rate of drying is accelerated by low relative humidity, high air temperature and good air movement around the cut forage. The water loss is rapid initially but slows as drying progresses. Three quarter of the water may be lost in the first one fifth of drying time with the final drying rate being less than one hundredth of the initial rate (Pitt and Shaver, 1990). Since leaves of cut herbages lose water more rapidly than stems, mechanical conditioning (crushing or crimping) also can reduce the time required for curing. However, this effect generally is greater for legumes than grasses. Reducing the curing time is of critical importance in haymaking. Field curing is conducted during bright sunny weather. Spread the cut fodder right after the dew has dried up and the sun begins to shine. However, this may result in bleaching of the forage and loss of leaves due to shattering One can also dry fodder on a fence, tripod or any similar structure, but this method is rather laborious. One must heap up the hay at night to protect it from dew, and spread it twice on the first day of drying and at least once during the succeeding days. To avoid this, drying can be done in barns by passing hot air through the forage. Although artificial drying produces hay of good quality, it is expensive and beyond the reach of small and marginal farmer but can be attempted on a community basis in areas where there is a need, and the necessary facilities.

4. Control of Moisture content

The freshly cut crop has moisture content of 70-80% but this is too much for hay making, so the following moisture content has to be kept for making good hay.

(i) For loose hay:25-28% moisture

(ii) For bundled hay:22-25% moisture

(iii) For chopped hay:19-22% moisture.

In order to judge whether hay is properly cured, two thumb rule methods are followed.

i. The scrape method

With nail try to scrape the epidermis from the stem of the plant. If the epidermis can be peeled from the stem, it is not sufficiently cured.

ii. The twist method

A small amount of the hay is twisted in the hands. If the stem breaks a little and no plant juice comes out from the twisted stems, the hay can be assumed to have been cured well.

5. Store on a well-drained site

Well dried hay should be stored properly for feeding to livestock. Hay should not be stored until it has dried completely, since wet or moist hay (in addition to encouraging the growth of moulds) may also ferment. The heat produced during fermentation can be a fire risk. Hay must be stored in a dry environment. Hay can also be stored by creating hay stacks. These may be created in a field near the source, or close to where the hay will be required later in the year. Stacks may be covered by plastic sheets to keep out rain. The surface layer of a stack may also be "thatched", in the same manner as a thatched roof to a house. Hay of similar quality should be stored together. High-quality hay should be stored inside or should be protected from the weather and raised off the ground on old tires, poles, or crushed rock. Break direct contact between damp soil and hay using rock, tires or poles.

Methods of Hay Making

There are four popular methods of hay making. They are as follows:

A. Field curing

B. Mow curing

C. Artificial drying

D. Pelleting.

A. Field curing

As the name indicates, cut plants are cured in the field itself, to make hay. The various steps in this process are (1) Cutting the crop (2) Swath curing, (3) Raking (4) Cocking (5) Bailing and storing.

1. Mowing

The first step in harvesting hay is mowing the standing crop and laying it in windrows. The maturity of the grass is the determining factor for starting the first field of the season. The grass should be in the early vegetative stage, and not headed out, with enough growth to make mowing worthwhile. A good strategy is to mow just before or right after a rain, because of the likelihood of good weather for the next few days. Some people like to mow early in the morning to gain almost a day of drying time. Others prefer to mow at the end of the day when the grass is drier, in an attempt to increase the energy level of the forage by capturing some of the sugars that migrate up the plant stem during the day. To cut the crop any type of mower, or hand cutting may be done. Mowing Equipment may be sickle mower, sickle haybine or rotatry disk mower

Sickle mower

One of the earliest methods of mechanically mowing hay was the sickle mower. A sickle is a hand-held agricultural tool with a variously curved blade typically used for harvesting grain crops or cutting succulent forage chiefly for feeding livestock. The blade may be semicircular, straight or slightly curved; it is usually inserted into a wooden handle by a tang; some have the blade attached to a metal shaft with a wooden handle. Sickles come in a wide range of forms, and two types of blade edge: serrated or smooth. While both can be used for cutting either green grass or mature cereals, it is the serrated sickle that still dominates the duty of reaping. It is still used today, typically in small haying operations or for clipping pastures. These generally have a 5'-7' cutter bar. Sickles are designed for cutting cereals, but are poorly adapted to mowing hay, and very slow and laborious in comparison to the scythe. It is a cheap, simple tool and is still very widely used all over the world. Sickle-cut hay is usually handled in bundles from the time of cutting, and is often carried to the homestead for final drying. The individual amounts are usually small. The sickle mower has a few disadvantages:

- Clogs easily with thick grass and hay that is bent over.
- Has a limited swath width.
- Can't crimp or squeeze the grass stalks (to improve drying) in one operation.

Scythe

Scythe is a manual implement for cutting grass, having a long handle held with both hands and a curved sharpened blade that moves in a plane parallel to the ground.

Two quite different types of blades are used: ground blades (sharpened on a grindstone) and hammered blades (sharpened with a hammer and small anvil). The blade, of mild steel, has a reinforced back, the end of which forms the tang for connection to the snead. Its length for mowing cereals and herbage is 75 - 100 cm (shorter blades are used for brush-cutting); longer blades require more skill in operation and a stone-free field surface. The width of the blade at the heel is 10 cm or less for mowing, but may be wider for dealing with thick-stemmed material.

Grass, especially thin stands of natural hay, is much easier to mow with a scythe in the morning when the dew is still on it. Mowing with a scythe is a skilled job, and very different from cutting with a sickle. Forage crops are mown with wide cuts, advancing in a straight line and cutting away from the standing crop. A scythe cuts about 2½ times the length of the blade, so a 75 cm one will mow a swath 175 cm wide if the crop is not too heavy, with the width reduced in a denser crop. On good, level, stone-free ground, about 500 m^2 can be mown per hour.

Sickle haybine

As the next development in haymaking, the sickle haybine greatly improved the mowing process. It uses the sickle bar technology, but has a front reel that lifts up bent-over stalks and allows it to be pulled at a faster speed. In addition, the sickle haybine has a built-in crimper that cracks open the grass stems to condition the hay for quicker drying. It also deposits the hay out the back into a gathered swath. The sickle bar haybine is still used on many medium-sized farms.

Rotary disk mower

Rotatry disc mowers, which have a cutter-bar fitted with two to six rapidly rotating discs, are a more modern tool and stand up much better to situations of poor field levelling and coarse herbage. It consists of several rotary disks equipped with small knives at the bottom that spin at a very high speed. These mow through thick hay well and can be pulled at higher speeds than conventional style mowers. The small blades are inexpensive and can be replaced when dull. The disk mower can also be outfitted with a crimper, which deposits the hay in a gathered swath. Rotary disk mowers have a good turning radius, and come in models that will mow from 6' swaths to 10'-16'. Disc mowers can operate at up to 16 km/hr, but 10 to 13 km/hr is more practical. Their working rate is 0.5 to 1.5 ha/hr per metre of cut. The rotary disk mower has become the industry standard in recent years.

2. **Tedding**

Once the hay starts to dry, it needs to be worked to promote curing. Tedding, the next step in haymaking, fluffs up the cut hay and allows the air and sun to contact the under-surfaces to promote drying. Hay tedders are generally wide units with several orbital wheels that lift the hay as they turn. Some styles have horizontal bars with teeth on a spinning reel. Some people ted immediately after mowing to spread out the swath. Hay mowed early in the morning could be tedded that afternoon, as long as the mowed swath is dry on the top surface. It may require a second tedding the next day to speed up the drying process. Too much tedding can shatter leaves of alfalfa or clover, lowering the quality of the hay. Proper tedding can be the key to timely haymaking.

Hay is dried much more rapidly in the swath than window. Therefore maximum advantage of swath curing may be taken to speed up the operation. But after a certain degree of curing, there will be shattering and bleaching of leaves lowering the nutritive value of hay considerably. The forage should be left to cure in the swath until it is wilted sufficiently. But before that there is danger of shattering and loss of carotene due to bleaching action of sun.

No definite time can be assigned for swath curing as it is dependent on climatic variables and the nature and density of the crop. Roughly the moisture at this time will be 40 percent.

3. **Raking**

Once the hay has been tedded and is nearly dry, it is ready to rake. Raking turns the hay one more time to dry the bottom and forms it into a windrow ready to be baled. After wilting forage to about 40 percent moisture in the swath it is rolled into small loose fluffy cylindrical bundles known as windrows. The windrows shouldn't be rolled too tightly, as this creates a roping effect that prevents the hay from drying properly and causes it to clog as it enters the baler. This is preferably done by the side delivery rake or with hand. It is better to do raking in the morning as dew makes the hay a little more tough and prevent shattering. As a rule of thumb, wait to rake hay until after the dew has dried and the sun nears its peak, or around 11:00 am. If possible, let the raked hay sit for an hour or two before baling to allow more drying time.

Hay makers have several types of rakes available. The old style side-delivery rake, which can be ground-driven, is pulled at an angle to the tractor and has several rows of teeth on multiple reels that roll the hay into a windrow. A pin-wheel rake consists of four or five large wheels with teeth mounted on them. The wheels turn by friction as the rake is pulled over the field. Newer types of rakes have a series of rotary wheels that pull the hay together into a windrow. Some models of tedders are designed so they can be converted to a raking mode. Most haymakers rake the outside swath in first, going counter-clockwise around the perimeter of the field. Then they reverse directions and roll the first windrow to the outside and continue in clockwise circles around the field. Some prefer to re-rake the first and second windrows inwards after the rest of the hay is baled, to allow additional drying and to make an easier passage with the baler

4. **Cocking**

Where there is danger of re-wetting, which is very damaging, there is an intermediate stage where the partially-dried herbage is put into small heaps - "cocks" - to dry further with less danger of rain affecting the whole mass. This is the process of making windrows into bigger heaps after hay had been cured partially in the windrows. Cocks are even protected with hay caps where rain and storm are expected. Under sub-optimal conditions, the cocks may be built on wooden tripods to improve ventilation, or the hay may be put on drying racks - all labour-costly operations. They are also useful in hot dry areas where the best quality hay is made by putting it up in cocks, which reduces both shattering and weathering and produces a green, leafy hay high in carotene. Cocks and large windrows are also used in hot, dry areas to avoid shattering and sun bleaching. Due to labour shortage this step is discarded in many advanced countries.

5. **Bailing**

A bale is a large quantity of hay tied together tightly. There are three main types of bales that can be made for feeding livestock: small square bales, large square bales and large round bales. Large round bales are quite popular among many livestock producers, especially those who raise beef cattle, and the largest type of bale a producer can make to feed livestock. Science and art converge in haymaking with the critical decision of when to start baling. Baling hay too early will trap moisture in the bale and result in spoilage. Baling too dry will cause leaves to shatter and break, lowering hay quality. It takes close visual observation and handling of clumps of hay from several windrows to "feel" if it is ready. Hay ready to bale will have no bunches of green grass. It will have a brittle, crisp feel. To test its readiness, grab a clump of hay and hold one end of the clump in each hand. Pull vigorously with both hands. If the stalks break on the first pull, the hay is ready to bale. If it takes several pulls, it's still too green.

A baler is an implement pulled by a tractor, that packs the hay into a bale of either round or square then tyed to hold the hay togathor. There are many different types of balers consisting of small square balers, large square balers, and round balers. Small square balers are the most common type and are usually done on the small operations. The large sqaure balers are a growing product because of efficiency and the ability to save labor. The round baler is a very efficient way to bale hay but the bales are very had to haul in mass quantities because they are hard to stack on semi trailers. Pick up bailing directly from windrow is the midst auto-mated system where the baler attached to a tractor picks up in the form of windrows and bales it. Stationary balers, are also available where hay has to be carried to the baler. The machine must be adjusted according to the type and moisture content of the hay, which will vary according to the time of day and position in the field, so continual operator attention is essential. Moist hay gives heavy bales, which may burst the strings if too dense and/or too large. Light, stemmy hay gives light bales with loose strings. Where such machines are not available hay may be stored as loose bundles.

6. **Storage**

Hay can be kept for long periods if properly made and correctly stored; in contrast, it can deteriorate rapidly and even be lost by careless storage. The aim in storing hay is to keep it dry and to protect it from wastage due to rots, pests, stray livestock, fire or wind. Storage areas, whether stack-yards or for bales, should have a dry foundation throughout the year, not only at the season of storage. They should be accessible all year round to the type of transport used, and protected from stray livestock and any fire risk. Where possible, it is better to store close to the point of use rather than in the field. Barn storage is ideal, of course, but is not always feasible.

Stacks of loose material are the traditional method of storing large quantities of hay and, apart from barn storage, were the principal system on large as well as small farms until the introduction of the pick-up baler revolutionized commercial haymaking. Once the hay has been field-dried to a suitable stage, it is carted to the stack-yard and built into stacks. The hay must be carefully built on to carts in the field and taken off in the same order as it was loaded. Stacking is a skilled job and the quality and safety of the harvest depends heavily on the skill of the builder, especially in regions of moderate to high rainfall. Stacks may be cylindrical or rectangular in section, depending on the quantities involved and the condition of the hay. Often they are built on a base of loose stone for aeration, and in humid-temperate zones a central wooden tripod can be put at the centre of cylindrical stacks for aeration. Skerman and Riveros (1989) recommend stacking on a base of hardwood poles, but this will depend on availability and price. When building the stack, the centre must be tightly packed, otherwise the stack will sag as it settles and rain may penetrate. The sides should slope outwards slightly to the eaves and the head thereafter rise gently to a point or ridge. Ropes or nets are used to protect the stacks against wind. Thatching with coarse grass or straw is common in areas where rain is expected during storage. This should not be done immediately, because it is necessary to allow dissipation of heat and moisture following stacking.

The yard must be kept tidy throughout stacking and storage, and rats should be eliminated if possible. Apart from the damage they may do, rodents in stacks in warm areas attract snakes, which can be distracting to the labour when the hay has to be taken out for livestock feeding. Termites are another possible cause of damage. In areas subject to high winds, windbreaks, such as earth walls, have to be constructed, else it is not possible to store hay in the open. In severe cases, the stack may be no higher than the windbreak. Storage of hay and straw in the forks of trees is widespread, especially for straw, in order to keep it out of reach of wandering livestock. House-top storage is also very common on smallholdings in semi-arid areas where flat-roofed houses are traditional; the house roof is often used as a drying area for other crops, as well as hay. In areas of negligible precipitation, this is an excellent storage method, but where there is rain the hay will be damaged by weathering. One reason for using the roof top is to keep the hay out of the reach of livestock around the homestead.

Once in store, they must be kept dry and protected from stock, damp and fire. This may require further thatching or covering of stacks (where hay is stored loose) and keeping the storage area clear of loose material. In traditional systems, stacked hay is usually taken from the stack to a barn or store before distribution to stock. If the store cannot hold a whole stack, the remainder left outdoors must be protected from rain and wind by heavy plastic sheeting or a tarpaulin, otherwise the hay will be spoilt, weathered, and possibly blown away. Carting of loose hay and straw to livestock, housed or in the field, must also be done so as to avoid loss. Baled fodder is simpler to handle because the bales are practical units for distribution.

B. Mow curing

Barn drying is the other name for mow curing. Mow curing refers to the practice of curing partially dried hay inside the barns in mows. Heated or unheated air is blown on to the mows until the moisture is reduced to 20-25 percent.

Swath curing is completed in the field itself and when the moisture is 35-40%, it is taken in the barns and placed on; the mows. It takes 7-14 days on the mows with unheated air to cure the hay fully, with heated air it takes less time. Generally the hay produced in this manner will be greener and leafier and of a higher quality than field cured hay.

C. Dehydration

Dehydration is also known as artificial drying. This is the process of chopping freshly cut or wilted fodder and drying it in artificial driers.

D. Pelleting

The hay is made into pellets by a pelleting machine. This has the following advantages, (i) simplifying hay making, (ii) lessening transportation and storage space from one-third to one-fifth, (iii) reducing labour (iv) increasing the density resulting in higher dry matter intake, (v) decreasing nutrient losses (vi) making automatic feeding of hay possible, and (vii) eliminating dust.

However, making hay under this method requires costly equipments and therefore the small farmers cannot afford it. Due to inclement weather, hay which is wilted to 50% moisture content have to be stacked sometimes. This forage ferment and reduces in nutritive value due to excessive heat generated. These hay will turn brown and hence the name brown hay. It is poorer in nutritive value and is only prepared when there is no other alternative.

Hay Making under Rural Condition

Under rural farming conditions the following procedures helps in making hay. The drying of green fodder can be done in the following ways.

1. On the ground

As estimated quantity of 100 q/ha cut green material can be utilized. The initial dry matter content may be 25 percent. About 0.2 hectare of land is required. The harvesting, mowing, tedding of the fodder is done immediately for intensive drying. Several tedding follow depending on drying and weather conditions. Wind-rowing is done. If necessary the wind-row is turned. Then the hay is stored in stacks.

2. On frames

This can be conveniently used to directly cut material to prevent leaf losses. 500-600 kg green material per unit can be dried. Five units of two sides are convenient. This necessitates 20 poles of 200 cm length and 4.5 cm in diameter.

In this method, the frame is set up and green fodder is packed evenly, beginning with the lowest step, it is carefully placed, on last layer on the top of the frame to prevent rain water from running into the fodder.

3. **On small fences**

This can be operated in small handy units. Six units of 400 -500 kg green fodder each can be conveniently operated. This method requires 40 poles of 250 cm in length and 4-6 cm in diameter and both ends pointed.

Seven pieces of plastic string or impregnated sisal- string of 35m in length, each is also required. The technique is to punch holes in the ground at a depth of 40 cm, and at a distance 120 centimeters. Then the poles are set, the lowest string is stretched first, 50 cm above the ground with special knot. Then the green fodder is packed evenly. Then above that second and third strings are stretched. Pack up finally a portion on each pole to serve as a roof.

4. **On tripods**

This method is ideal for one-day pre-wilted lucerne, berseem etc. Four units can be conveniently operated with 600 kg green material per unit of 300 kg per-wilted material. The requirements are as follows: 12 poles of 250 cm in length and 6-8 cm in diameter. 12 poles of 200 cm in length and 6-8 cm in diameter and 12 iron sticks, 30 cm long, to be fixed in the upright poles. To make the unit, set up the tripod and fix the horizontal poles on the sticks pack up the fodder evenly working round the tripod.

5. **On poles**

In this method, labour and material requirements are high. But, handy small units provide very good drying conditions. Each unit can utilize 40-50 kg green fodder. About 60 units can be profitably deployed at a time. Poles with 250 cm length and 6 cm in diameter both ends pointed with 4 holes are required. In the holes, small sticks of 60 cm length and 2.3 cm in diameter are inserted.

To make the unit, punch holes in the ground; minimum 40 cm deep and set the poles. First, we should fix the lowest cross stick and pack up the fodder evenly. Then, fix the second stick and pack up simultaneously. Finally it should be packed as a roof.

Types of Hay

Depending on the crop from which hay is made, they may be classified as legume hay, non-legume hay, grains crop hays and mixed hay. The different types of hay is described here in detail.

1. **Legume hay**

Legume hay are of special value to dairy cattle. The hay contains higher digestible nutrients and the digestible protein is of superior quality. They are rich sources of carotene, vitamin A, E and K. It is considerably rich in calcium and is highly palatable. Legume hay are made out of leguminous plants like lucerne, berseem, cowpea, groundnut, and kudzu. The digestible crude protein content range from 6.7-14% in these types of hay. Legume hay are usually fed to young l calves in calf starter rations.

Some of the perennial fodder legumes like *Centrosema pubescens, Glycine javanica, Atylosia scarabaloides* and *Dolichos lab lab* for feeding to livestock either as such or converting them into hay.

2. **Non-legume hay**

The out-turn per hectare is more but the hay made is less palatable and nutritious compared to legume hays. The non-leguminous hay contains less proteins, minerals and vitamins. This type of hay may be made from either crops or grasses.

3. **Grain Crop hay**

Grain crop hay is made from oats and barley, harvested at milk stage. It has low protein and mineral content but is rich in carbohydrates. Oat is the only graminaceous forage crop of winter. For hay making in February -March the oat needs to be kept for about 120-144 hours in shade and 96-120 hours in sun to dry. Total digestible nutrients of oats is 59.8% and the crude protein is 6.1 per cent.

4. **Mixed hays**

Mixed hay is made from a mixed crop of leguminous and non-leguminous plants. Such a crop is generally cut early as they do not come to the flowering stage simultaneously. The nutritive value of mixed hay will depend on the proportion of leguminous and non-leguminous plants.

Feeding Hay to Livestock

Hay is roughage, and is usually far from a complete feed. Even from high-potential crops with modern machinery, the resultant hay is often inadequate to provide a production diet for cattle. High quality legume hays made in dry warm climates are usually of high feeding value, but grass hays in more humid climates and from tropical herbage are of lower quality. Only some legume hays can provide enough protein to supplement other roughages. In stall-fed production systems, as in dairying and fattening, the role of hay and straw is to provide bulk in the ration and, according to quality, some energy and protein, but will usually be supplemented with concentrates and, in season, green fodder. Mature work oxen can subsist on very coarse roughages, although they may need some concentrates during hard work. Growing and milking animals are a very different case, as their needs for protein and minerals are high (and in proportion to growth or the milk yield).

Poor feeding techniques are a common source of wastage. Hay is often fed in the field by simply distributing it on the ground, and this can lead to heavy wastage through trampling and contamination. Feeding off the ground, apart from being wasteful, is a health hazard. In some situations, feeding on the ground may be unavoidable, but racks should be used wherever possible. Hard standings help avoid contamination, but are expensive. Hay and straw are often fed to overwintering stock for maintenance, so bales are distributed in the field, leading to waste and treading. When dry fodder has to be fed loose in the field, waste will be less if it is distributed little by little so that the stock eat it all and wait for more. Long hay and straw should be fed from racks or mangers to reduce wastage and the racks should be designed for the type of animal to be fed. In arid areas, troughs are often built from mud; elsewhere wood, metal or masonry are used. Where buffalo are being fed, troughs must be particularly robust, otherwise the stock will smash them. A lot of wastage from racks and troughs can be avoided if they are lined or covered with weld-mesh to restrict the amount which animals can pull out at one time. The selective feeding habits of sheep are a problem, but mesh over the trough helps. Portable racks, on wheels, are used for field feeding, and should be sited either on hard standing or on a well-drained part of the field, and shifted regularly to reduce damage to the sward. Standard bales must be opened for feeding. Particular care should be taken with wire-tied bales, to ensure that no pieces of wire are left in the feed or around the feeding area since, if ingested, they may cause serious illness or death. For housed stock, the racks are often fitted above the trough or manger so that any feed which falls from the rack is held in the trough. Simple feeding systems can be made from locally available materials: wood, brick and stone - or even dried mud in arid places - and are associated with the improvement and intensification of animal production everywher.

Where the residues are removed from the field at harvest, as in intensive irrigated systems to allow field preparation for the succeeding crop, residues can be fed to stock elsewhere, allowing about twice the amount they might require, permitting them to select (and reject); in this system the residues are not chaffed. Coarse, rejected, residues need not be wasted, for they can thereafter be collected, urea treated and then fed to stock. Selective feeding on straw does not necessarily assure an adequate diet, although it is a considerable improvement on non-selective feeding, so, for productive stock, supplementation should be done necessarily.

Losses During Hay Making

Nutrient losses in hay starts at the moment it is cut. The dry matter losses can be 35-40% of the cutting. Nutrient potential loss can be 20-60%. This is major economic loss for the hay grower as the hay crop will be graded lower. Hay losses in yield and quality start in the field. Just because the "hay" has been cut doesn't mean plant respiration stops. Chemical changes resulting in losses of nutrients arise during the drying process. These losses are as follows:

Respiration Losses

Even after harvesting, plant cells alive and functioning until the moisture content falls below 35 to 40 percent. Sugars are the primary plant carbohydrates lost during storage respiration. Under good drying conditions, respiration loss accounts for 2 to 8 percent dry matter whereas the loss can be upto 16% under poor drying conditions (Klinner and Shepperson, 1975). Respiration loss from hay curing is proportional to the moisture content and temperature (Collins and Moore, 1995). During extended storage periods, microbial and fungal respiration reduces oxygen levels and produce carbon dioxide, water, and heat, causing the reduction in dry matter. Dry matter losses could exceed 10 percent if moisture levels are between 20 and 30 percent

(Klinner and Shepperson, 1975). Harvesting hay when good drying weather is expected will reduce respiration losses considerably.

Management practices that shorten drying time resulting in reduced respiration and harvest losses include: (a) cutting early in the day to maximize solar drying, (b) cutting when anticipated weather will allow for relative humidity of the air to be below the equilibrium humidity of the forage, (c) the use of mechanical or chemical conditioning to reduce the cuticular resistance to water escape and (d) maximizing hay exposure to wind and sunlight by creating wide and thin windrows.

Losses by Leaching

The uncertainty of weather conditions always makes haymaking difficult. Rain lowers the quality of hay through leaching of water-soluble carbohydrates and prolonging respiration losses. If hay is almost cured and is exposed to heavy and prolonged rains, especially when it is in the field, loss of soluble carbohydrate, minerals, water soluble vitamins and nitrogenous constituents through leaching. The extent of leaching loss is influenced by several factors including type of forage, stage of maturity, moisture content at the time of rainfall, amount of rainfall, frequency of rain and mowing/conditioning treatments (Bolsen, 1985). Alfalfa harvested in the bud stage undergoes more extensive leaching loss than hay harvested in full bloom presumably because the amount of soluble nutrient decrease as the alfalfa plant matures. Leaching by rain can cause up to 20 percent nutrient loss. Rain may prolong the enzyme action within the cell cells, thus causing greater losses of soluble nutrients, and may also encourage the growth of molds. Rainfall-induced losses are greater for drier hay. In research at Purdue University. one inch of artificial rain on field-cured hay reduced total digestible nutrient (TDN) content 5 percentage points.

Losses by Shattering

During the drying process the leaves loose moisture more rapidly than the stems, so becoming brittle and easily shattered by handling. Alfalfa leaves dry down 2 to 5 times faster than stems and as plant moisture decreases to below 30%, leaves become extremely brittle (Shaeffer and Clark, 1976).

The loss due to shattering of leaves in hay making is of importance, especially in the case of legumes. The leaves at the hay stage are richer in digestible nutrients then the stems and hence losses by shattering of leaves decrease the nutritive value of hay. Leaf loss is nutritionally and economically important because alfalfa leaves comprise approximately 50% of the crop dry matter and contain over 70%of the plant protein, 65% of the digestible energy and 90% of the plant carotene (Bohstedt, 1944). Hay moisture content is the largest single factor contributing to leaf loss. When the hay becomes too dry and brittle and losses become excessive. To avoid these losses, hay should never be over dried or handled during warm periods of the hay. Hay stored at moisture content above 15 percent has much less leaf loss than hay stored below 15 percent moisture. If hay becomes too dry, stop baling and resume in the evening or morning when the leaf moisture level increases. This dew-moistened hay can be stored at a slightly higher moisture level than when it was drying down because dew moisture in the hay is more easily released during curing than internal moisture.

The upper moisture limit to prevent mold growth and other hay deterioration for large round alfalfa bales is typically 18 to 20 percent. Hay stored above 25 percent moisture will usually spoil unless chemical preservatives are added to the hay. Hay that is put up too wet also increases the chance of fire, especially when stored indoors.

Losses by Bleaching

During the process of drying, much of the carotene, a precursor of vitamin A is lost with bleaching. In general, the carotene content of freshly cured hay is proportional to the greenness. With severe bleaching, more than 90% carotene may be lost.

Storage Losses

Everytime the hay is handled, some more nutrition is lost. Storage losses are directly related to microbial growth and to subsequesnt heating. Excessive heat damage can reduce protein and energy digestibility of the hay. The amount of storage losses are directly related to several factors *viz.* moisture content at baling and the time of storage, storage conditions, environmental conditions (relative humidity, air temperature, and air movement), forage species and the epiphytic microbial populations present on the hay (Tomes, 1989). Hay that is stored at moisture contents greater than 20 percent can develop mold and lose dry matter and quality to bacterial degradation.

The biological activity in hay doesn't necessarily terminate at bailing, especially if bailing is done at higher moisture (20-30%) to reduce leaf shatter losses. Hay does not become static until it reaches about 12% moisture and the equilibrium humidity is below 65% at which time most fungi will not grow (Tomes, 1989). If hay is stored at moisture contents between 15 and 20 percent and not protected by a preservative or inoculants, hay undergo some elevation in temperature in the first 2 to 3 weeks after baling. This heat buildup is referred to as "sweating" which leads to measurable losses of 4 to 5 percent in hay DM. The extent of heating largely depends on (1) the moisture of the hay, (2) the density and size of the bale, (3) the rate of bale dry- down and (4) the epiphytic microbial populations present on the hay. The first temperature peak will generally occur within a few days and can be the result of aerobic bacterial growth, fungal growth and/or plant respiration. If oxygen and a favorable moisture level are available, microorganisms begin to multiply; generating heat up to 130 to 140° F. Heat resistant fungi and bacteria are known to be active when temperatures are between 130 and 150° F. The rise in temperature tends to kill most microorganisms resulting in the gradual decline in internal bale temperatures. The initial heating in hay baled at lower moistures typically drives off moisture. However, in higher moisture bales, the hay moisture combines with water generated in the respiration process, allowing for unusually prolonged conditions that prove optimum for bacterial and fungal growth. A single, continuous temperature rise is often observed when hay is baled at higher moistures (over 30%) as a result of favorable growth conditions for selective aerobic bacteria and fungi. The hay may, however, undergo several heating cycles during the next few weeks as various populations of microorganisms increase and decrease. Eventually the temperature will stabilize near ambient temperature (Prather, 1988). Heating above 175° F results in the eventual death of all microbes, however, the previously generated heat can stimulate heat producing chemical reactions that further increase temperatures. Oxidation of reactive compounds may ultimately cause the temperature to rise to an ignition point of 448 to 527° F. If enough oxygen is present under these conditions, spontaneous combustion will occur resulting in fire (Martin, 1980). Excessive heat damage will reduce protein and energy digestibility of the hay.

Aerobic fungi are the primary microbes responsible for the breakdown of complex carbohydrates and subsequent generation of heat (Martin, 1980; Tomes, 1989). The primary nutritional losses that occur during storage are due to microbial growth and the subsequent heating. Excessive heat damage can reduce protein and energy digestibility of the hay. Heat damaged protein is measured by determining the nitrogen content of the fiber in tests such as the Acid Detergent Insoluble Nitrogen (ADIN) analysis. Under normal conditions, less than 5% of the total nitrogen should be bound to the fiber fraction. It is generally felt that excessive heat damage has occurred when ADIN approaches 10% or more of the total nitrogen (Ricketts *et al.*, 1982; Shelford, 1983). Brandt *et al.* (1984) found that when the moisture content of large round bales increased from 15% to 27%, the ADIN as a percent of total nitrogen increased resulting in a decrease in protein digestibility from 71 to 53 percent. Dry matter digestibility was also reduced by 5% indicating that losses in digestible energy also occurred in the hay.

Mold growth in improperly cured hay can adversely affect palatability and feed intake, although less than 5% of the molds commonly found in hay produce any mycotoxin (Tomes, 1989). Mohanty *et al.* (1969) showed that feeding moldy alfalfa hay resulted in significantly lower dry matter intake, reduced weight gains and poorer feed conversion compared to feeding mold-free hay. He also reported a 25-30% decrease in the feed value of the alfalfa. Weathering also contributes significantly to storage losses when hay is stored outside, although hay stored inside at normal moisture levels of below 20% can still experience 5-10% dry matter losses (Martin, 1980). Most weather deterioration is limited to the outside layer of the bale and at the soil surface. These losses are dependent upon: a) amount of rain fall, b) length of storage, c) storage site and d) ability of the bale to shed water. Legumes generally experience more weather losses than grass hay because they do not form as tight a weather-resistant thatch. Petritz (1988) indicate that hay, even when stored inside, can lose as much as 1% of the dry weight for each percentage unit of moisture loss. This loss is related to the above-mentioned metabolic activity in the hay during the final stages of drying. Hay baled at 20% moisture will likely lose 5-8% of its dry weight by the time it cures to 12% moisture. When hay is stored outside, additional dry weight loss occurs due to weather damage. The total loss, including the unavoidable 5-8% loss and that due to weathering can be as high as 40 percent. Martin (1980) suggested that losses with large round bales stored outside on the ground are about three times greater than those protected by inside storage. Furthermore, those storage losses can range from 10-42% of the original dry matter. Belyea *et al.* (1985) studied large round, alfalfa bale storage and feeding losses finding

40% total losses for bales stored outside with no cover, 20% losses for bales stored outside but covered and 15% for bales with inside storage. Rain penetrated 4-10 inches into the uncovered bales stored outside contributing to feeding losses up to 25%.

Benefits of hay making

Hay is the most commonly used stored feed on most farms. Hay making has following advantages:

1. It can be kept for long periods of time with little loss of nutrients if protected from weather;
2. A large number of crops can be successfully used for hay production;
3. It can be produced and fed in small or large amounts;
4. It can be harvested, stored and fed by hand or the production and feeding can be completely mechanized;
5. Hay can supply most nutrients needed by many classes of livestock.

References

Bareeba, F.B. 1992. Foarage conservation and utilization. *Nomadic Peoples*. 31: 91-96.

Belyea, R.L., Martz, F. A and Bell, S. 1985. Storage and feeding losses of large round bales. *J. Dairy Sci.*, 68: 3371-3375.

Bohstedt, G. 1944. Nutritional Values of Hay and Silage as Affected by Harvesting, Processing and Storage. *Agricultural Engineering*. p.337.

Bolsen, K.K. 1985. Personal Communication. Kansas State University, Manhattan, USA

Brandt, B., M. Nelson, T., Klopfenstein, and Britton, B. 1984. Influence of Silage Inoculant on the Yield and Preservation of Influence of Silage Inoculant on the Yield and Preservation of Alfalfa Baled at Three Moisture Levels. Nebraska Beef Cattle Report.

Collins, M. and Moore, K.J. 1995. Post harvesting processing of forages. In: *Forages - the Science of Grassland Agriculture.* Vol. II. (ed.) R.F. Barnes, D.A.Miller, and C.J. Nelson. Iowa State Univ. Press, Ames, Iowa.

Klinner, W.E. and Shepperson, G. 1975. The State of Haymaking Technology: A Review. *J. Br. Grassland Society*. p. 303.

Martin, N.P. 1980. Harvesting and storage of quality hay. American Forage and Grassland Council Proceedings. p. 177.

Mohanty, G.P., Jorgensen, N.A., Owens, M.J. and Voelker, H.H. 1969. Effect of Molding on the Feeding Value and Digestibility of Alfalfa Hay. J.Dairy Sci., 50:990.

Pit, R. E and Shaver, R. D . 1990. Processes in the Preservation of Hay and Silage. In Proc. Dairy Feeding Systems Symposium (NRAES-38), Harrisburg, Pennsylvania.

Prather, T.G. 1988. Hay Fires - Prevention and Control. University of Tennessee Agricultural Extension Publication PB 1306.

Petritz, D. 1988. How Much Can You Pay For Indoor Hay Storage. Hay & Forage Grower. March, 1988. p. 19.

Ricketts, R., Belyea, R., Sewell, H. and Garner, G. 1982. Understanding and Interpreting Feed Analysis Reports. University of Missouri Extension Publication.

Sheaffer, C.C and Clark. N.A. 1 976. Effects of Organic Preservatives on the Quality of Aerobically Stored High Moisture Baled Hay. *Agron. J.*, 67:660.

Shelford, J.A. 1983. Personal Communication. Department of Animal Science, University of British Columbia.

Skerman, P.J. and Riveros, F. 1989. Tropical Grosses. FAO Plant Production and Protection Series, No. 23. Flood and Agriculture Organization of the United Nations, Rome

Suttie, J. M. 2000. Hay and Straw Conservation - For Small-Scale Farming and Pastoral . *FAO Plant Production and Protection Series No. 29.* Food and Agriculture

Organization of the United Nations, Rome. http://www.fao.org/docrep/005/ X7660E/x7660e00.htm

Tomes, N.J. 1989. Personal Conversation. Research Microbiologist, Microbial Genetics, West Des Moines, IA.

CHAPTER 15

Hay Quality

Hay quality really means feed value. The ultimate test of hay quality is animal performance. Quality can be considered satisfactory when animals consuming the hay give the desired performance.

Characteristics of Good Quality Hay

1. Hay should be nutritious therefore prepared form plants cut at an appropriate stage of maturity when it has the maximum nutrients.
2. Good hay should be leafy. The leaves are generally richer in proteins, minerals and vitamins then other parts of the plant.
3. Hay should be green in colour. The green colour indicates the amount of carotene which is precursor of vitamin A. The green colour should be preserved by minimizing the bleaching, leaching arid fermentation losses.
4. Hay should be soft and pliable.
5. Hay should be free from weeds and stubbles.
6. Hay should be free from dust and molds.
7. The moisture content in hay shouldn't exceed 15%.
8. It should have the smell of and aroma of characteristics of the crop from which it is made.

Factors Affecting Hay Quality

Two principal methods can be used to determine forage quality. The first is to have a chemical analysis. Where a chemical analysis is not available, the second method of hay quality evaluation involves a visual inspection of hay and a judgement based on certain physical characteristics of forage. Hay is judged visually for relative feeding value and overall desirability as a feed. The factors known to influence hay quality and animal performance include the following:

1. Stage of maturity at harvest
2. Leafiness
3. Color
4. Foreign material
5. Odor and condition.

1. **Stage of Maturity**

Stage of maturity refers to a plant's stage of development at the time it is harvested. Of all the factors affecting hay quality, stage of maturity at the time of harvest is the important factor influencing quality. The younger the plant, better the quality. As legumes and grasses advance from the vegetative to reproductive (seed) stage, they become higher in fiber and lignin content and lower in protein content, digestibility, and acceptability to livestock. Determining the maturity of legumes and grasses is easy before harvesting but becomes more difficult after cutting and baling. Weathering or sun bleaching after cutting or the delay of normal development of legume flowers due to cool, cloudy weather, especially with first cutting, further complicate the determination of maturity. For top quality hay, legumes should be cut at the 20 percent bloom stage. Grasses should be cut in the fully headed to early bloom stage. Most forage will have a 20 percent loss in TDN (total digestible nutrients) and a 40 percent loss in protein by a delay of only 10 days past the most desirable stage of harvest.

2. **Leafiness**

Leafiness is an excellent indicator of hay quality. This refers to the ratio of leaves to stems present and is also related to the stage of maturity, especially in grass hays. The younger the plant, the higher the proportion of leaves. As a grass matures, stems increase, thus decreasing the quality of the forage. Leafiness in legumes is particularly critical because legumes lose their leaves during curing and handling more readily than grasses. A high percentage of leaves also indicate good harvest and handling methods. In general, hay with high leaf content has a higher percentage of minerals and vitamins and a greater energy value than hay with few leaves. As legumes and non-leguminous forages advance in maturity, the stems mass increases, lower leaves fall from the plant, and the leaf-to-stem mass ratio decreases. The method of curing of hay, the method of handling it from field to storage, and the weather conditions during curing and baling also influence leafiness. Leaves shattered from the stems and loose in the bales may be wasted when the hay is fed. To preserve leafiness, hay must be cut early and carefully cured and handled.

3. **Colour**

Color can be a definitive characteristic of hay, which tells more about the curing process of the hay than its quality. The most desirable hay color is the bright green of the immature crop in the field. Hays having this bright green color were most likely cut at an early, desirable stage of maturity and were well and rapidly cured, with no damage from rain, molds, or overheating during storage. Green hay is rich in carotene, whereas, straw-colored or brown hay is very poor in carotene. Yellow color is often a result of sun bleaching and does not seriously reduce quality. Dark brown or black is often an indicator that the hay was exposed to rain or high humidity and is usually accompanied by a distinctive musty odor. Overall, slight discolorations from sun bleaching, dew or moderate fermentation are not as serious as the loss of green color from maturity, rain damage or excessive heating or fermentation. Slight discolorations from sun bleaching, dew, or moderate fermentation are not as serious as the loss of green color from maturity, rain damage, or excessive fermentation or heating.

Golden yellow coloured hay/ Sun-bleached hay

Sun-bleached hay is light golden yellow in colour. Sun bleaching reduces palatability and carotene content. In general, the amount of carotene in hay is directly proportional to its degree of greenness.

Yellow coloured hay

Yellowing, especially in grass hay, usually indicates that the plants were over-matured when cut. Grasses with fully ripe seeds usually have yellowish-brown stems and heads and many brown leaves. Yellowing due to maturing can be distinguished from sun bleaching because all of the plants, rather than just those on the outside, have the same yellowish color.

Dark brown coloured or Rain damaged hay

Hay that has been exposed to rain or to heavy dews or fog has a characteristic dark brown or black appearance. Rain falling on hay between the time of cutting and baling can leach out soluble nutrients and reactivate plant respiration. Hard rain can also shatter leaves, especially if the rain occurs when the hay is nearly dry. The stems of hay that has been sun-bleached or discolored by rain are usually harsh and brittle. Avoiding rain damage is a goal in haymaking but delayed cutting must be balanced against the lower feeding value of late cut hay.

Brown coloured hay/ Heat damaged hay

Brown hay indicates heating from microbial (mold) growth and fermentation. This results when the hay is stored at too high a moisture content. This hay has a distinctive musty, moldy odor and when the bale is opened the flakes are often caked and show visible mold. Molds consume nutrients in the hay, particularly sugars and starches, producing carbon dioxide and water. Extensive heating, as with brown hay, results in considerable loss of dry matter, digestible protein, and energy and destroys much of the carotene and other vitamins. Slight discoloration from sun bleaching, dew, or moderate fermentation is not as serious as the loss of green from maturity, rain damage, or excessive fermentation.

Color	Problem	Quality
Green	None	Usually good, this is hay cut at the best time for feeding
Light golden yellow on outside of bale	Sun bleaching	Decreases palatability and carotene
Yellow throug-hout bale	Over-mature when cut	Decrease in palatability
Dark brown or black hay	Exposed to rain, heavy dews, or fog	Decrease in nutrient content due to leaching, leaf shattering may be present
Brown hay	Mold growth, hay not dry when baled	Musty, moldy odor, flakes may show mold, loss of dry matter, digestible protein, energy, carotene and other vitamins, don't use this hay

4. Foreign Materials

Foreign materials in hay can be divided into injurious and non injurious categories. Injurious foreign material is material that will harm the animal if eaten. This includes poisonous plants and matter such as wire or nails. Non-injurious foreign material is matter that is commonly wasted in feeding operations but is not harmful to livestock if eaten. This includes weeds, stubble, chaff, and sticks. Weeds are the most common non-injurious foreign material found in hay. Livestock do not relish most weeds and if they are eaten they usually have little or no feed value. Weed seeds usually pass undigested through livestock and when the manure is spread on land it becomes a source of weed infestation.

5. Smell

The smell of new mown hay is the standard by which hay odor is judged. A fresh aroma, freedom from mold and relatively high carotene content add to its palatability and feed value. Mildew, mustiness, or a putrefied (rotten) odor result from weather damage or insufficient drying before baling and indicate lower quality hay. Odour problems usually result in lower acceptability by livestock. Hay should be free from dust, mold, insect and disease damage.

Hay fire and hay quality:

Hay fires are a major problem in stored hay. Freshly cut forage is not dead; respiration (the burning of plant sugars to produce energy) continues in plant cells and a small amount of heat is released in the bale. Many producers refer to this elevation in bale temperature as "sweating" or "going through a heat." In hay that is baled at the proper moisture concentration, plant cell respiration has slowed dramatically and will eventually cease. The heat generated by plant cell respiration in hay bales is normal and generally of little consequence. However, if bale moisture levels are too high (greater than 20 percent), the heat and moisture will provide a suitable environment for the growth and multiplication of mesophilic (warm temperature) bacteria that are present on forage crops. The respiration of mesophilic bacteria releases additional heat in the bale and interior bale temperatures can reach 130° to 140°F. At this temperature range, most mesophilic bacteria die and interior bale temperatures start to decline. This cycle of heating and cooling may occur several times during the weeks after baling as the microbial population increases and decreases. However, the maximum temperature decreases during each subsequent cycle. The interior bale temperature will eventually stabilize near the ambient temperature. Hay that has sustained these heat cycles has lost much of its quality as a feeding source, but is unlikely to catch fire. Baled hay becomes a potential fire hazard when the interior bale temperature does not cool after the first heating cycle. This occurs when the respiratory heat created by the mesophilic bacteria provides an environment favorable for the growth and multiplication of thermophilic (heat loving) bacteria. The thermophilic organisms multiply and the heat produced by their respiration can raise the interior bale temperature to 170°F before microbial activity ceases. The thermophilic bacteria and their respiration heat convert the hay to a form similar to a carbon sponge with microscopic pores. This damaged material combines readily with oxygen at high temperatures and can self ignite in the presence of oxygen.

A variety of factors are believed to influence hay heating and the risk of spontaneous combustion. These include moisture (or dry matter) content of the forage, crop or pasture species, air temperature and humidity, bale size, compaction and stacking. Other factors can also contribute to increased risk of hay heating and fire including variation in plan water soluble carbohydrate (WSC) levels and types, along with changes in waxy leaf cuticle (surface) due to moisture stress, uneven drying due to uneven crops, weeds or machinery used. As a general guide, hay which is baled at more than 20% moisture is at risk of heating or going mouldy which will affect nutritional quality of the feed. If hay is baled at 25 to 35% moisture (75 to 65% dry matter), then spontaneous combustion is likely to occur. There is a very narrow range of moisture levels suitable for making large bale hay. If the forage is baled too wet it will heat. If it is baled too dry, leaf shatter and field losses during baling increase and can exceed 25% if the hay is less than 12% moisture.Risk of heating is affected by bale size, density and how it is stored. Smaller bales have a relatively large surface area, so can dissipate heat. Under dry conditions hay in small bales and lower density round bales will continue to lose some moisture after baling. However, once stacked in a shed the effective size of the lot increases and the comparative advantage of smaller bales to dissipate heat is lost. Bale stacking configuration will also affect heat build-up. Tightly stacked square (rectangular) bales will not allow air movement and heat will build up. Some arrangements of round bales may allow heat to dissipate out of the stack, however damage is still possible in the middle of the bale (Bruhn and Kogel, 1985).

Heating of hay without burning is often beneficial, because it increases palatability of the hay. However, feed value of the hay is likely to be reduced. Heating causes changes inthe chemical structure of the hay which reduce its feed quality (Coblentz *et al.*, 2010). Chemical bonds form between carbohydrate and protein, reducing metabolisable energy content and protein availability.

Maillard or browning reaction

When excessive heating occurs (>38°C) amino acids, which are the building blocks of protein, react with sugars to become less degradable or less digestible. This reaction also occurs when cooking to brown meat or caramelise onions. The effect on hay is similar, with heated hay becoming caramelised and quite palatable to livestock. However, heating means the hay has lost energy and the Maillard Reaction means it has lost digestible protein. Browning reactions also produce more heat which can eventually lead to spontaneous combustion. Heating of hay following baling result in loss of total dry matter, energy and digestible protein of the hay.

There are many methods to prevent hay fire. Hay moisture concentration has a major effect on the microbial activity that can lead to hay fires. Therefore, hay should be cured to the proper moisture concentration prior to baling. Weather conditions during hay curing have the greatest influence on achieving proper moisture concentration. Ideal hay-curing weather is slightly windy with a relative humidity of 50 percent or less. Hay should not be baled in the early morning because its moisture concentration increases overnight when the air is humid. Therefore, carefully monitor the weather forecasts before scheduling hay-making operations. Using specialized hay equipment can help reduce hay moisture concentration by increasing crop drying rates, especially during good hay-curing weather. Conditioning equipment crimps or abrades crop stems to break up the waxy cuticle layer surrounding the stem. This process allows plant moisture to evaporate and speeds up the crop drying rate. Tedders fluff, spread, or move windrows to improve air movement through the crop. Windrow turning equipment lifts, inverts, and fluffs windrows. As a result, the bottom side of the windrow is then exposed to the sun, which speeds up drying during good hay-curing weather (Grisso *et al.*, 2002).

Hay preservatives can reduce the risk of hay fires because they inhibit or reduce the growth of bacteria in moist hay. Propionic acid is a liquid that is applied to hay during baling and is the most effective preservative available. However, a tank and a spray application system must be added to the baler to handle the preservative. Furthermore, spray nozzles must be spaced so that the chemical is distributed over all the forage as it enters the baling chamber. The amount of propionic acid that must be applied depends upon the hay moisture concentration. Small bales with a moisture concentration between 20 percent and 25 percent should be treated with about 0.5 percent propionic acid on an as baled basis. The application rate should be increased to 1 percent for hay with a moisture concentration between 25 percent and 30 percent. Propionic acid and other hay preservatives may not work for hay with a moisture concentration greater than 30 percent. The two main disadvantages of using propionic acid are its corrosiveness and the cost of the equipment necessary to apply the preservative.

Hay that is baled at the proper moisture concentration can become a potential fire hazard if it becomes wet during storage. Hay barns should be weather tight and have adequate drainage so water will not enter the barn and damage hay during storms. Hay that is temporarily stored outside should be covered with plastic or some other waterproof material to prevent rain penetration. Uncovered bales should be arranged so that air can circulate freely between bales to promote drying. Protect bales from ground moisture by placing them on a bed of gravel, old tires, poles, or pallets.

The temperature of hay that has been baled at a high moisture concentration should be checked twice a day for six weeks after baling.

Use a probe and thermometer to accurately determine the temperature inside a stack of hay. A simple temperature probe can be made in the farm shop from a 10-foot piece of 3/4-inch diameter iron pipe. Drill eight 3/16-inch diameter holes about three inches from one end then hammer that end of the pipe together to form a sharp edge . Commercial temperature probes are available, but are often too short to monitor the maximum interior temperature zone within a hay stack. Check hay temperature in the center of the stacked hay. The easiest way to do this is from the top of the stack. Do not walk directly on the stacked hay; pockets may have already burned out under the hay surface. Place boards, plywood, or a ladder on the hay and walk on those instead. This will spread the weight of the person monitoring the temperature over a larger area and help keep him from falling into burned out cavities. Be sure to use a lifeline in case the hay surface collapses into a fire pocket. A second person, standing safely away from the hay, should hold the other end of the rope to pull the person monitoring the temperature out in case the hay surface collapses into a fire pocket.

Quality Characteristics of Hay

Excellent quality

Legumes cut in the late bud to early bloom stage for alfalfa or 20 percent bloom stage for clovers; grasses cut in the boot stage. Hay is bright green, leafy, and free from mold or mustiness. This feed is high in carotene, protein, minerals, and energy and low in fiber.

Good quality

Legumes cut by half bloom; grasses cut in the early heading stage. Good quality hay is leafy, has good green color and is free from mold or mustiness. Hay that was rained on after partial curing can fall into this category if it was cut early. The nutritional value is not as high as excellent quality hay but good results can be obtained if it is properly supplemented (Banerjee, 1998).

Fair quality

Legumes or grasses cut at full bloom. Fair quality hay lacks greenness and is stemmy and low in carotene, minerals, protein, and energy and high in fiber.

Poor quality

Any legume or grass cut after full bloom. Poor quality hay is stemmy and lacks leaves. It may be severely weather damaged, bleached, musty, or moldy.

References

Banerjee, G.C. 1998. Conservation of forage crops. In: *Feeds and Principles of Animal Nutrition.*Oxford and IBH Publishing Co. Pvt. Ltd., New Delhi, India. pp. 35-51.

Bruhn, H.D. and Kogel, R.G. 1985. You can avoid silo and haymow fires. Hoard's Dairyman June 10, 1985: 653.

Coblentz, W.K, Hoffman, PC, and Martin, N.P. 2010. Effects of spontaneous heating on forage protein fractions and in situ disappearance kinetics of crude protein for alfalfa-orchardgrass hays packaged in large round bales. *Journal of Dairy Science* 93: 1148–1169.

Griffiths, N, Piltz, J, Clayton E, Meyer R and Richards, S. 2010. Reduced feed quality of hay following heating. Proceedings NSW Grassland Society Annual Conference, Dubbo, July 2010. pp. 127–130

Grisso, R., Smith, R., and Cundiff, J. 2002. Management Tips for Round Bale Hay Harvesting, Moving, and Storage, VCE publication 442-454. Virginia Cooperative Extension, Virginia Tech, Blacksburg, Va.

McBeth, LJ, Coffey, KP, Coblentz, WK, Turner, JE, Scarbrough, DA, Bailey, CR, and Stivarius, MR 2001. Impact of heating-degree-day accumulation during bermudagrass hay storage on nutrient utilization by lambs. *Journal of Animal Science.* 79, 2698–2703.

Murphy, D.J., and Arble, W.C. 1986. "Extinguishing Silo Fires." NRAES-18. Natural Resources, Agricultural, and Engineering Service (NRAES): Ithaca, N.Y.

Nadeau, J. 2007. Hay Analysis: Its Importance and Interpretation. Fact Sheet. Department of Animal Science, University of Connecticut

Orloff , S.B. 1997. Hay curing, baling and storage. In Intermountain Alfalfa Management (Eds S.B. Orloff and H.L. Carlson). Publication 3366, Agriculture and Natural Resources. University of California.

CHAPTER 16

Hay Preservatives

Storage of hay at higher moisture contents may lead to some heating due to the activity of aerobic micro-organisms and possibly some plant respiration. The warm, moist conditions in hay will provide the ideal environment for growth of spoilage bacteria, e.g. Bacilli and yeasts, moulds and fungi. These organisms utilize the energy and protein of the hay and can lead to a substantial increase in their respective populations. Their action leads to the following reaction:-

Hay (plant sugars) + oxygen $\rightarrow$ Carbon dioxide + water + heat

The resultant heating causes a reaction between the carbohydrates and proteins rendering both less digestible as temperatures continue to increase. The best opportunity to reduce hay-harvesting losses is by shortening curing time. A combination of mechanical conditioning, chemical conditioning and high-moisture baling help accomplish this goal (Collins, 1988; Mahanna, 1994).

Hay preservatives are chemical or biological agents that inhibit or reduce the growth and activity of these aerobic micro-organisms (bacteria, yeast and moulds) in the hay so stops production of the water and heat. Moulding and heating is avoided, so negligible nutritive value is lost. Most hay preservatives do not improve the nutritional quality of the forage, but do prevent the decline in quality caused by the micro-organisms. Hay preservatives allow the conservation of hay at higher moisture (of up to 35 percent) thereby reducing drying times and leaf shattering losses associated with handling of dry forage. The use of hay preservatives permits greater flexibility in hay making operations.

Types of hay preservatives

Hay preservatives have been classed into several types. These are:

- Organic acids and their salts
- Bacterial inoculants
- Sulphur based preservatives
- Ammonia-based additives
- Common salt
- Enzymes
- Antioxidants
- Nutrients.

I. Organic acids and their salts

Organic acids have been the most widely accepted hay preservatives. Organic acids act as fungicides (Knapp *et al.*, 1976; Sheaffer and Clark, 1976) by producing an acid environment which is not conducive to mould, yeast or bacterial growth. Propionic and acetic acid being naturally occurring acids in the rumen, are safe for all types of livestock, including horses. Other organic acid, such as citric acid is sometimes also included, but propionic acid is the most effective as a mould inhibitor. Moulds greatly reduce the value of dry hay. Moulds consume hay nutrients and cause dry matter losses, as well as producing toxins that are detrimental to animal health. Mouldy, dusty hay contains spores that can cause respiratory problems, particularly with horses. Mould growth can even result in hay fires from spontaneous combustion.

The propionic acid is highly effective against mould growth whilst acetic acid is more effective against bacteria and yeast. Propionic and propionic/acetic acid (vinegar) blends are good for preventing molds and heating damage in high-moisture hays. How effective the organic acids are depends on the rate of application and moisture content of the hay. The fectiveness is largely dependent upon the application rate of active ingredients and the moisture content of the hay. Recommended application rates of actual acid for small square bales generally range from 0.5%- 1% for 20-25% moisture hay up to 1.5%

acid for 31-35% moisture hay. These rates are appropriate for propionic acid alone or mixtures of propionic acid and acetic acid (80:20 percent). In the market, solutions will vary from 10 to 100% propionic acid. The uses of very dilute products are not recommended because larger volumes of water are applied to the crop. Walgenbach (1986) suggests that products containing a high percentage of propionic acid have successfully preserved "wet" hay when applied at rates of 1% or more of the as is weight of the hay. However, when considering the economics of applying upwards of 20 pounds per ton of hay, data suggested that propionic acid was only economical when used to reduce losses incurred from impending weather damage (Sheaffer and Martin, 1979). Walgenbach (1989) indicates that the expected results of propionic acid on hay depends upon the complex relationships between: (a) the level of inhibitory, free propionic acid, (b) the level of the non-inhibitory, ionized, propionate form of the acid, (c) the buffering capacity of the crop and (d) the hay microbial populations, some of which are capable of metabolizing the protective, free acid form. Propionic acid products attempt to inhibit microflora growth on moist hay by striving for high levels of the inhibitory, free acid form of propionic acid (CH_3-CH_2-COOH) and lowered levels of the non-inhibitory, ionized form of the acid (CH_3-CH_2-COO^-). Late season mold growth has been a problem sometimes associated with acid hay products. Walgenbach (1989) suggests that the mold inhibition of acid products is greatly reduced after long periods of hay storage. The acid eventually dissipates during storage while the moisture remains in the hay. This sets up ideal conditions for mold growth especially in hay removed from storage late in the feeding season.

These products are corrosive on machinery and can be dangerous for operators to use in their pure form. To overcome these problems, "buffered" acids, some times referred to as "neutralized or pH balanced" acids, have been developed and commonly include salts of propionic, acetic and formic acids. In recent years, there has been much interest in "buffered" propionic acid mixes that have a pH of about 5.5 to 6. These products are based on ammonium or sodium salts of propionic acid. Buffered acid products consisting of compounds such as ammonium propionate act in a manner similar to normal propionic acid. These compounds disassociate into: a) the free acid and propionate forms, depending upon the surrounding pH, and b) ammonia based compounds that can also exhibit microflora inhibitory properties. The newer buffered products are much less volatile and corrosive than the original unbuffered acid, making them much easier to use. The reduced volatility prolongs the activity in the bale with less odour. Another chemical used is the "acid-salt" or sodium diacetate (food preservative). It acts by raising the acetic acid (vinegar) levels in the moist material. The main problem in using this product is the vinegar taste and smell. Buffered products, although apparently equal in performance, generally cost higher than normal acid products (Thomas, 1989).

Certain precautions should be observed in using organic acid-based preservatives. Protective clothing should be used in mixing or transferring the material, and water should be available at all times to flush affected areas if an accident occurs. Equipment surfaces and applicator systems should be thoroughly washed and flushed immediately after use to prevent excessive corrosion. Treated hay should be stored under protective cover because the preservatives can be leached or diluted easily on outer surfaces by rainfall. Preserved hay should not be stored in contact with normally field-cured hay, since the drier hay can absorb moisture, making mold development possible. Although hays containing moisture levels higher than 35 percent moisture can be effectively preserved with these materials, the practice is not recommended because of preservative costs and difficulty of handling wet bales.

Hay treated properly with buffered propionic and other organic acid products is safe to feed to livestock, including horses. Propionic and acetic acids are produced as an end product of digestion by bacteria in the rumen of cattle and sheep, and the caecum and colon of horses. The animal then uses these organic acids as a nutrient in the digestion process and metabolism.

II. **Microbial Inoculants**

Some silage inoculants developed for silage use are also effective for hay preservation. There are three basic groups of bacterial or microbial inoculants products used as hay preservatives:

- Fermentation enhancers
- Aerobic spoilage inhibitors
- Antibiotic-producing, bacterial inoculants

Fermentation enhancers

Some silage inoculants developed as fermentation enhancers are also effective for hay preservation. Most contain lactic acid producing bacteria that compete with yeasts and mould forming organisms aiming to maintain forage quality. Lactobacilli would produce acid, reduce pH and thus reduce mold growth. Commonly used lactic acid bacteria are *Lactobacillus, Pediococcus* and *Streptococcus.* When using this class of inoculants on hay, the plant sugars must be high (substrate for the bacteria) to provide the best protection against moulds and yeasts. For pasture plants this usually occurs when they are vegetative, in mid to late spring.

Aerobic Spoilage Inhibitors

These recently developed inoculants use *Lactobacillus buchneri,* a group of bacteria which restrict the growth of spoilage type organisms such as yeasts and moulds. Their mode of action is thought to be due to secondary metabolites called "buchnericides" and may reduce the need for paddock curing unless extremely wet. Recent research indicates that they are usually much more effective than the fermentation enhancer group.

Antibiotic-producing bacterial inoculants

Another group of bacteria is *Bacillus amyloliquefaciens.* These bacteria produce antimicrobial compounds which inhibit, sometimes stopping, the growth and reproduction of moulds and other spoilage micro-organisms. These bacteria also increase the rate of water loss which makes hay less conducive to spoilage organisms. The increased drying rate reduces the rate and amount of plant respiration, resulting in less nutrient loss and reduces the period for population growth of spoilage organisms. The digestion trial showed that lambs fed inoculated hay gained equal to those fed dry hay and numerically outperformed the untreated high-moisture hay group in average daily gain (Emanuele et al., 1992).

III. Sulphur based preservatives

Sulphur based preservatives are another product that control of microbial proliferation. Sulphur compounds scavenge oxygen. In fodder, they create an environment within the hay bale that is unconducive to microbial growth stemming mould and yeast development. The application rate at 0.8-1.0 kg per tonne is much lower then with acids. Sulphur compounds are not acidic, and as such are generally fairly user friendly.

IV. Ammonia-based preservatives

Anhydrous ammonia

Anhydrous ammonia has fungicidal properties and has been used successfully in the preservation of high-moisture hays. Use of 1 percent anhydrous ammonia (dry matter basis) to hay containing up to 30 per cent moisture has been shown to reduce storage dry matter losses and prevent heating and mold development. The ammonia apparently sterilizes the hay, killing organisms that cause hay to mold. USDA researchers have effectively baled alfalfa hay at 30% moisture content with the addition of 60 pounds of ammonia per ton. However, unless covered with plastic to retain ammonia, the preserving effects of the ammonia will be only temporary (Walgenbach, 1986). Ammonia treated hay should be stored in small stacks, covered with plastic and the anhydrous permitted to equilibrate with the hay for at least three weeks (Rohweder *et al.,* 1983). The most common method for treating hay with anhydrous ammonia is to seal the hay in plastic and inject the ammonia into the sealed area. Several variations of these methods are observed which are dependent upon the amount of hay to be treated. Single large round bales are put into large plastic bags and sealed. A 3 to 4 foot length of black metal pipe attached to a hose from the ammonia source is inserted into the bag and one to three pounds of ammonia are applied for each 100 pounds dry matter weight of the bale. The pipe is then removed and tape is used to seal the hold in the plastic made by the pipe.

The addition of anhydrous ammonia also combines with the residual moisture in hay forming ammonium hydroxide that breaks the lignin-cellulose bonds in the cell walls of the forage (Davis, 1980). Increased crude protein content is an additional benefit of ammonia preservation. It also solubilizes some of the complex carbohydrates in the plant and swells plant fiber, thereby allowing for greater rumen microbial breakdown of the forage. The improvement in digestibility will enhance dry matter intakes of low quality forages by 15 to 20% due to an improvement in the digestive rate of passage (Kuhl, 1982). Anhydrous ammonia will also provide a source of non-protein nitrogen that may or may not be utilized by the animal depending upon nutrient demand and sources of ration protein (Sniffen and Chase, 1987).

However, this method of chemical preservation has not received wide acceptance because application of anhydrous ammonia to large volumes of hay is difficult and it is a hazardous chemical.

Urea

Dry urea could be used as an alternative to anhydrous ammonia in preserving high-moisture hays and increasing the crude protein content of poor-quality hays. Applying urea, which is then converted to ammonia by the bacteria, is simpler than applying ammonia gas. Urea offers advantages in that it lacks the volatility, corrosiveness and potential for user injury exhibited by ammonia. Alhadhrami *et al.* (1989) conducted research in Arizona on early bloom alfalfa harvested at approximately 25% moisture, treated with 2% or 4% urea and fed to mid-late lactation cows. Comparisons were made against a wet (31% moisture) and dry (11% moisture) control. Results showed visually less mold on the 4% urea treated bales and lower post harvest temperatures when compared to the 2% treatment or wet control. Feed intake and milk yield in lactating cows fed the hay was not significantly different between treatments. At the end of the 4-month trial, in vitro neutral detergent fiber digestibilities (48 hours) were significantly higher in the urea treated hays. Caution was exercised in feeding the 4% urea treated hay by dilution of 50% of the total forage intake with normal, untreated hay. The researchers concluded that feeding 100% of the forage as the 4% treated hay probably would have produced toxic effects even though 41% of the urea nitrogen had disappeared prior to the feeding trial. Relatively large amounts (5 to 7%) of urea applied during baling can be effective up to 30 per cent moisture. However, the treated hay must be covered tightly with plastic sheeting as soon as possible after baling.

V. Common salt

The use of salt (NaCl) on wet hay does have a biological basis in that sufficient concentrations will absorb free water on the surface of the hay and thereby inhibit microbial growth. Common salt has often been used between layers of moist hay to both preserve the hay and improve its palatability. However, it is ineffective as a preservative unless applied in such amounts as to be physiologically harmful to animals. The salt itself would be detrimental to microbial and mould growth, and being hygroscopic in nature, absorbs moisture in proximity to the salt granules. The placing of salt between layers of large rectangular bales will only be effective near the salt itself, possibly having an effect of 1 to 3 cm into the bale itself. The only way salt may be effective would be if the salt was spread throughout forage in the bale (Lacefield, 1987).

VI. Hay Desiccants/drying agents

Drying agents are the first technological advancement to hasten the field drying of forages since the advent of mechanical conditioners in the mid 1940's. Hay desiccants are used to reduce the length of time needed for hay drying. The chemical drying agent (potassium or sodium carbonate, sodium silicate and citric acid) are applied to the stem of the alfalfa plant at the time of mowing. These naturally-occurring salts reduce drying time by acting on the moisture-conserving, waxy-cutin layer of forage (Rohweder *et al.*, 1983). These chemicals are most effective when applied at 5-7 pounds of active ingredient in 30 gallons of water per acre. This allows the stem to dry faster and can reduce the interval between cutting and baling. Hay desiccants may decrease time needed for drying by one third.

Dessicants are less effective with: (a) heavy windrows that often occur at first cuttings, (b) uneven application or (c) rainy, humid conditions. It was observed that dessicants were less effective in humidities over 80% and that treated hay tended to rewet faster from dew exposure, however, the desiccant treated hay also dried faster when good drying conditions reappear (Rohweder *et al.*, 1983). Leaching losses can be greater with desiccant treated forage if hay is rained on while lying in the field. Hay desiccants work the best in good drying conditions and are less effective in poor drying conditions. Chemicals can be corrosive to cutting conditioning machinery and bales may be heavier. Leaching losses may be greater in desiccant-treated alfalfa if rained on before baling.

VII. Enzyme-based Preservatives

The enzymes used in preservatives are simply enzymes taken from inside bacterial inoculant cells. Their purpose in the industry is slightly different than bacteria inoculants in that their main goal is to promote plant cell breakdown and render the cellulose and starch found in the plant fiber (ADF & NDF) more accessible to desirable acid producing bacteria. While they do not directly prevent mold growth, enzymes will make nutrients available to desirable lactic acid bacteria (LAB), thereby increasing the desirable LAB. The greatest effects are often seen on more mature forages with higher fiber and moisture contents, or those forages with limited sugar contents.

Common types of enzymes used are Cellulase, Amylase and Protease, and unlike bacterial inoculants, enzymes can be stored for long periods of time. However this is dependant on the process in which they were derived. It is recommended to spray all inoculants and enzyme-based preservatives directly on the forage as being harvested to maximize application uniformity. In comparison to acids and ammonia, bacterial and enzyme-based preservatives are generally not as effective in preventing mold. However, they are less hazardous to handle, readily available, easy to apply, and they are relatively inexpensive.

Factors Affecting Hay Additive Performance

Consideration should be given to several factors that exert influence on the efficacy of any hay additive.

1. Moisture Content

Most hay additive products suggest an upper moisture limit for which the product is recommended. This is typically in the range of 25-35% moisture. Adhering to these moisture limits is probably the most important factor in assuring satisfactory performance of the product. When paying for technologies such as hay additives, it seems prudent to take the time and effort to accurately determine the conditions under which the product will best perform.

2. Application and Calibration

Applicator set-up and calibration are critical management steps in assuring that the correct amount of additive is uniformly distributed throughout the hay. Hay growers should be sure that information is provided with regards to: a) calibration, b) nozzle types, c) nozzle pressure, d) applicator positioning and e) active life of the product once put into the applicator.

3. Environmental Conditions

The specific environment that hay is exposed to can affect the performance of hay additives primarily due to the profile and activity of spoilage organisms that thrive on the hay. Conditions such as rain can also alter the microflora or leach soluble carbohydrates contributing to varied product performance. Hay growers should solicit research conducted in their specific climate when considering product adoption.

4. Storage and Rate of Drydown

When baling hay at higher moistures, it is presumed that the extra water must eventually migrate from the hay. If hay is baled at higher moisture rather than the typical 15% moisture, the removal of additional water will be required during the storage period. Hay will eventually stabilize at 12-15% moisture but the time required to reach this level depends upon many factors including: a) initial bale moisture, b) relative humidity, c) air temperature, d) air movement, and e) bale density.

Rate of drydown seems to exert an effect on the pattern of microflora growth with implications as to the quality of hay exposed to long term storage. Management practices should be adopted that maximize the rate of drydown such as: a) using well ventilated storage, b) stacking alternate layers at right angles to one another, c) leaving some air space between bales, d) stacking in several small piles to increase surface area, and e) not storing field cured hay next to wetter hay treated with a hay additive product.

5. Realistic Product Expectations

Hay growers should have realistic expectations of hay additives. Additives will not be a management "cure-all". Rather, they are a value-added product designed to enhance existing management practices. Additives can also be considered a "risk management" technology, allowing growers a management option in combating weather. There will be certain factors that growers and buyers will have to consider when using additives. Often bales will not display the "bright green" color typified by field-cured hay but rather a more "olive green" color. Growers will need to educate buyers to make purchase decisions based on laboratory analysis of the nutrient content rather than color. Producers may also have to content with factors such as loose strings due to shrink or unique stacking requirements; however, the added nutrient value coupled with the reduction in weather risk should be weighed against these minor inconveniences.

Benefits of Using Hay Preservatives

Benefits of using hay preservatives are that they:

1. Prevent dry matter and quality loss in storage due to bacterial, yeast and mould activity

2. Allow the safe storage of hay from slightly above target moisture levels up to 25% (or 30%) moisture depending on preservative type
3. Allow baling after a shorter curing period which reduces risk of rain damage and sun bleaching. It may also allow baling earlier in a season in certain areas, when fodder is less mature and nutritive value higher
4. Reduce dry matter and nutrient loss caused by leaf loss and shatter, microbial activity and moulds
5. Maintain hay colour (due to increased leaf retention) and often smells better
6. Reduce risk of spontaneous combustion
7. May increase animal intake
8. Animal and human health not affected due to lack of mould spores.

References

Alhadhrani, G, J.T., Huber, G.E., Higginbotham and Harper, J.M..1989. Nutritive Value of High Moisture Alfalfa Hay Preserved with Urea. *J. Dairy Sci.*, 72: 972-74.

Belyea, R.L., Martz, F. A and Bell, S. 1985. Storage and feeding losses of large round bales. *J. Dairy Sci.*, 68:3371-3375.

Brandt, B., Nelson, M. T., Klopfenstein, and Britton, B. 1984. Influence of Silage Inoculant on the Yield and Preservation of Influence of Silage Inoculant on the Yield and Preservation of Alfalfa Baled at Three Moisture Levels. Nebraska Beef Cattle Report, 1984.

Bohstedt, G. 1944. Nutritional Values of Hay and Silage as Affected by Harvesting, Processing and Storage. *Agricultural Engineering*. p.337.

Bolsen, K.K. 1985. Personal Communication. Kansas State University, Manhattan.

Collins, M. 1988. What Does Rain Damage Cost. *Hoard's Dairyman*. May 10, 1988. p. 433.

Davis, G.V. 1980. Ammoniation of Roughages and Grain. Formula Feed Conference. Garden City Branch of Kansas State Agricultural Experimental Stations. January 21-22, 1980.

Emanuele, S.M., Horton, G.M.J., Baldwin, J. , Lee, D. and Mahana,W.H. 1992. Effect of Microbial Inoculant on Quality of Alfalfa Hay Baled at High Moisture and Lamb Performance. *J. Dairy Sci.*. 75 (11): 3084–3090.

Klinner, W.E. and Shepperson, G. 1975. The State of Haymaking Technology: A Review. *J. Br. Grassland Society*. p. 303.

Kuhl, G. 1982. Ammonia Boosts Forage Quality. Dairy Herd Management. August, 1982. p. 16.

Klinner, W.E. and G. Shepperson. 1975. The State of Haymaking Technology, A Review. *J. Br. Grassland Society*. p. 303.

Knapp, W.R., Holt, D.A. and Lechtenberg, V.L. 1976. Propionic acid as a hay preservative. *Agron. J.* 68:120-123.

Lacefield, G.D. 1987. Why Salt Hay? *Hoard's Dairyman*. May 25, 1987. p. 463.

Martin, N.P. 1980. Harvesting and storage of quality hay. *Am. Forage and Grassland Council Proceedings*. p. 177.

Mahanna, W.C. 1994. Hay Additive Review: Where We've Been, Where We're Going. 24th National Alfalfa Symposium. Feb. 24-25, 1994, Springfield , IL.

Mohanty, G.P., N.A. Jorgensen, M.J. Owens and H.H. Voelker. 1969. Effect of Molding on the Feeding Value and Digestibility of Alfalfa Hay. *J.Dairy Sci.*, 50:990.

Prather, T.G. 1988. Hay Fires - Prevention and Control. University of Tennessee Agricultural Extension Publication PB 1306.

Petritz, D. 1988. How Much Can You Pay For Indoor Hay Storage. *Hay & Forage Grower*. March, 1988. p. 19.

Ricketts, R., Belyea, R. and Sewell, H.and G. Garner. 1982. Understanding and Interpreting Feed Analysis Reports. University of Missouri Extension Publication.

Rohweder, D.A., Collins, M., Finner, M and Walgenbach, R.1983. A Haymaking System To Help Make "Hay in a Day". University of Wisconsin Agricultural Extension Publication.

Rohweder, D.A. 1987. Intensive Cuttings Improve Alfalfa. June, 1987. Hoard's Dairyman.

Sheaffer, C.C. and N.A. Clark. 1976. Effects of Organic Preservatives on the Quality of Aerobically Stored High Moisture Baled Hay. *Agron.* J. 67:660.

Shelford, J.A. 1983. Personal Communication.Department of Animal Science, University of British Columbia.

Sniffen, C.J. and L.E. Chase. 1987. Field Application of the Degradable Protein System. Cornell University Advanced Dairy Nutrition Seminar for Agribusiness. September 8-11

Sheaffer, C.C. and N.P. Martin, 1979. Hay Preservation. University of Minnesota Agricultural Extension Publication no. 489.

Thomas, J.W. 1989. Personal Communication. Department of Animal Science. Michigan State University, East Lansin, USA.

Tomes, N.J. 1989. Personal Conversation. Research Microbiologist, Microbial Genetics, West Des Moines, IA.

Walgenbach, R.P. 1986. Hay Preservatives, Dessicants and Silage Preservatives Silage Preservatives -What We Know. Proc. 1986 Wisconsin Forage Council Meetings. January 28-29.

Walgenbach, R.P. 1989. Personal Conversation. U.S.D.A. Forage Research Center, Madison, Wisconsin, USA.

CHAPTER 17

Haylage

Haylage or baleage or round bale silage, is a somewhat newer method of preserving forage. Haylage is simply forage that is baled at higher moisture content than dry hay and then stored in a sealed plastic wrap. Because of the high moisture level and air-tight environment, the forage ferments and is preserved by acid production during fermentation. The higher pH values associated with baleage are related to a slower fermentation process. During the respiration stage of the ensiling process the oxygen that is trapped in storage is rapidly consumed by aerobic bacteria. Anaerobic bacteria, which survive in the absence of oxygen, begin to grow and multiply in the fermentation stage and convert the plant sugars into organic acids-mainly lactic acid and acetic acid. With production of acids, the pH of the silage is reduced from an original level of 7.0 to a final ideal pH of 4–5. Typical large bale haylage has a pH of 4.5–5.5. Fermentation ceases when bacteria growth is stopped by the accumulation of acids. The haylage will then remain at a stable pH with no bacterial growth and can be preserved for a long time, providing there is no exposure to air. Harvesting forages with a long fiber usually results in a slow release of the plant nutrients necessary for the growth of acid producing bacteria. This generally results in less fermentation in the baleage than in chopped silage.

Producing hay in the spring is a difficult task because of the high moisture content in the forages and cooler temperatures that extend the drying time. Making haylage out of harvested legumes and grasses is an alternate option for storing and processing forage. Forage can also be stored as baled silage as it provides advantages including high quality, flexibility in handling and feeding systems, limited capital expenditures and low losses during production and storage. Haylage can be made from any crop that is traditionally stored as hay.

How is Haylage Made?

Ensiling forages as baleage requires 40 to 60% moisture (with a targeted average of 50%). This is an intermediate level between hay and silage production. The forage is cut as if for hay-making but is baled at 50-60 percent moisture rather than at 18- 20 percent moisture. Baling at the proper moisture content is the single most important variable. At 40 to 60 percent moisture level, fermentation is more favourable in terms of sugar content than in wetter material and seepage losses are minimized. Ensiling material with too little moisture increases the risk of heating due to increased oxygen penetration, which will lead to mould growth and a marked depression in the protein digestibility. In addition, haylage can heat spontaneously if oxygen becomes available. Fires have been reported when the moisture in the haylage is near 40 percent. More often, the heating will form indigestible products which lower protein and energy values. A haylage with tobacco- brown or black color and a caramelized odor has undergone some spontaneous heating. Baling haylage with too much moisture reduces the feed quality of the forage and reduces the amount of dry matter stored per bag, greatly increasing storage cost.

Haylage is the most sensitive to moisture variation of the ensiled forages. Reductions in moisture content necessary for production of haylage are accomplished by conditioning (mowing, windrowing and drying for four to 24 hours) to reduce moisture levels to 40 to 60 percent. The drying time will vary on the climate, the type of forage, and the stage of cut. Considerably longer wilting periods might be necessary depending on the forage moisture and conditions. After reduction of moisture upto desired level, baling is done with a standard baler, and bales are wrapped with plastic shrink wrap. This is usually done with a large machine designed for this purpose. Storage of the bales is done either in a protected shelter, or on clean, smooth ground with no sharp stubble or rocks underneath to prevent puncture. Holes in the wrapper will allow additional air into the bale, causing it to mold or rot. If there is any hole in the bagged bales, patching should be done as soon as possible. Rodents can chew through the plastic wrap or bag, which will greatly increase storage losses. Spray the perimeter of the stack to kill weeds that harbor rodents and insects. The bales should not be covered with an extra layer of plastic because it may make an ideal nesting site for rodents. Shady area, preferably on a north facing slope is selected to store haylage to avoid temperature fluctuations that can degrade both the haylage and the plastic.

Storage of Haylage

Successful storage depends on many factors. The storage site should be cleared of stubble and sharp objects. Some people lay a piece of plastic on the ground prior to placing the bales. Rodents can chew through the plastic wrap or bag which will greatly increase storage losses. Spray the perimeter of the stack to kill weeds which harbor rodents and insects. Find a shady area, preferably on a north facing slope, to avoid temperature fluctuations which can degrade both the haylage and the plastic. If you find holes in the bagged bales, patch them as soon as possible, because wind causes loose plastic to bellow out, providing an air exchange which usually spoils most of the outer layer of the bale. Bags are rarely reusable because of minor pinholes. Ensiling forage as baleage can be a management strategy for high-nitrate grass. An average of 50 percent reduction in nitrate in ensiled forages due to the de-nitrification process was observed.

Feeding Haylage

Feeding haylage is similar to feeding large round bales of hay in that conventional feeding rings can be used. With the high investment in wrapping bales, it is essential to control feeding losses. Acids will be produced by fermentation in the wrapper, so a sour, fermented odor is expected, but dark or brown discoloration, obvious mold, or other signs of decay will mean the haylage may no longer be suitable for livestock feed, particularly for horses.

Haylage can be safely fed to cattle, sheep, and goats. It is not recommended for horses because of the risk of mold. If the haylage is improperly harvested or the plastic is damaged during storage, mold and mycotoxins can form in the bale. This can be toxic to horses and haylage should therefore be avoided if possible. The problem with feeding haylage to goats or sheep lies in reducing waste, as these small ruminants can get inside a typical ring feeder. When feeding individually wrapped haylage bales to any species, it is best to feed to a sufficient number of animals so they can eat an entire bale within one or two days. Once a stack is opened, deterioration of the haylage commences and continues at a rate that varies with air temperature. Eventually, mould becomes visible on the face of the bales and gradually will penetrate the bales. Therefore, once bale is opened; it should be entirely fed out within 1 week in summer; within 2 weeks in spring; and within 4 weeks in winter.

This method has certain advantages and disadvantages over other forage harvesting and preservation systems

Advantages of haylage

- Decreased curing time needed from cutting to baling, which makes weather less of a factor in forage harvesting.
- Potential for more timely harvest of large quantities of forage.
- Requires one- half to one-third the drying time of hay.
- Usually, haylage is higher in energy and protein because the cutting date is not dependant on the weather (less mature plants with higher carbohydrate and protein content and with less indigestible fibre).
- Decreased need for mechanical handling and time curing to dry the forage reduces the loss of leaves, the most digestible part of the plant.
- Potential for higher feed quality bale through leaf preservation and possible nitrate reduction.
- Dust free and more paltable than hay.

Disadvantages of haylage

- Increased harvest cost per bale vs. conventional cured hay.
- Disposal of used plastic wrap.
- More likely to spoil as compared to silage in traditional silos.
- Individually bagged haylage cannot be moved with the normal bale spears. A grasping unit is required to prevent damage to the integrity of the bags. If the bag or plastic wrap is damaged, oxygen enters and spoilage occurs.
- Risk of forage spoilage if integrity of wrap is not maintained. Birds and rodents can puncture plastic and holes must be covered.
- Handling problems can occur because the bags/bales are almost

twice the weight of dry hay (due to the 50% moisture level).

- Transportation of bales is limited due to cost of moving high-moisture bales.
- Storage life is shorter than that of chopped silage or dry hay.
- There is a greater danger of botulism when feeding round-bale silage or bagged silage than when feeding dry hay

References

Bates, D.B., Kunkle, W.E., Dawson, T. E. 1989. Round bale silage- a forage harvesting alternative. In: 38th Annual Florida Beef Cattle Short Course Proceedings, May 3-5. Univ. of Florida, Gainesville, Florida.pp. 65.

Battle G.H., Jackson, S.G and Baker, J.P. 1987. Acceptability and digestibility of preservative-treated hay by horses. *Nutrition Reports International*. 37 (1):83-89.

Hintz H. F, Lowe, J.E and Miller, W.F. 1983. Studies on the feeding of hay treated with mixture of propionic and acetic acid to horses. In: Proceddings 8th Equine Nutrition and Physiologica Scociety Symposium. University of Kentucky, Lexington, Ithaca.

Lawrence, L.M., Jaster, E.H., Wischover, L., Moore, K.J and Hintz, H. F. 1987. Acceptability of alfalfa hay treated with an organic acid preservative for horses. *Canadadian Journal of Animal Sciences*. 67(1): 217-220.

Lemus R. 2010. Utilizing Baleage as an Alternative Way to Maintaining Forage Quality Extension Forage Specialist. http://msucares.com/crops/forages/index.html

Martin, N. P., David R. M. and Paul J. W. 2004. Alfalfa: Hay, Haylage, Baleage, and Other Novel Products. Presented at the Idaho Alfalfa and Forage Conference 23-24 February 2004.

CHAPTER 18

Analytical Techniques for Hay and Silage

Analyzing feeds for nutrient content is essential for proper balancing of livestock rations. This is especially essential for those feeds that have undergone the fermentation process. Silage samples should be analysed for dry matter, protein fractions (degradable and undegradable protein, soluble protein, and heat damaged protein), ADF, NDF, mineral content and energy. In addition to these components, lactic, acetic, and butyric and propionic acids, pH, ammonia nitrogen, ethanol, and yeast and mold counts are also quantified.

pH estimation

The pH of silage measures the degree of acidity. The pH of silage is easy to determine. A silage sample can be prepared by vigorously shaking a small sample of silage with 100 ml of water for a minute or two. A pH-sensitive paper can be dipped into the liquid and will change color-based on the degree of acidity. This can be compared to a color chart and converted to an approximate pH reading. A pH of 7 is neutral and numbers less than 7 indicates acidic conditions. In general, a low pH means that more acid was produced. On the other hand, a high silage pH can have several causes. First, it may mean that the silage did not ferment well. Cold environmental conditions, lack of sufficient substrate for bacteria to make the acids, an undesirable fermentation (e.g. clostridia), or a low moisture content can lead to a high pH. A high pH could also be an indicator of silage that has spoiled due to exposure to air; badly deteriorated silages may have a pH greater than 7.5. A rapid production of acids inhibits bacteria while removal of air inhibits the growth of yeasts and molds.

Well preserved maize silage will have a pH in the range of 3.7 to 4.2

Dry matter estimation

On most animal operations, feed ingredients are provided to animals according to the dry weight of the feed. Dry matter is the percentage of the forage that is not water. Dry matter content is important because all animal requirements are made on a dry matter basis. Determining the DM content of feed provides a measure of the amount of a particular feed that is required to supply a set amount of nutrients to the animal. Increase or decrease in feed DM content result in over or under feeding of nutrients. It would be impossible to compare different forages without using the percent dry matter as a base line. Dry matter is also very important as the moisture content will give clues as to how a forage will preserve when stored as hay or silage.

Dry matter refers to material remaining after removal of water, and the moisture content reflects the amount of water present in the feed ingredient. Determination of DM is a relatively simple and quick process that can be easily done on a farming operation. The most common way that moisture is determined is through the evaporation of water from the feed, leaving only the dry contents behind.

Toluene Determination of dry matter

This procedure is recommended for determination of moisture in fermented feeds (silages) which contain high levels of volatile acids. Volatile acids and alcohols have low vaporization points and are lost when the sample is oven dried.

Principle

Water is distilled from sample and trapped under a layer of toluene.

Equipment

- 250 ml pyrex flask Bidwell-Sterling Moisture Receiver, calibrated to 0.01 ml by distilling known amounts of water into graduated column and estimating column of water to 0.01 ml.

- Clean tube and condenser with chromic acid cleaning mixture, rinse thoroughly with water, then alcohol, and dry to prevent undue amount of water from adhering to inner surface.
- 500 mm Liebig condenser
- Toluene distillation hood

Reagents: Toluene, reagent grade

Procedure

i. Weigh sufficient wet sample (at least 25g) to yield a minimum of 5 ml water, record weight (W_1) to nearest 1 mg, and transfer to 500 ml flask.

ii. Add sufficient toluene to cover sample completely.

iii. Immediately fill receiving tube with toluene, pouring it through top of condenser.

iv. Bring to boil and distill slowly, i.e 2 drops/sec, until most of the water passes over, then increase rate of distillation to about 4 drops/sec.

v. Distill 1 hr (or longer if necessary for wetter samples) to obtain clearing at the top of the condenser.

vi. When all water has apparently been distilled, wash down condenser by pouring toluene in at top and continue distillation for a short time (approximately 15 min) to ensure all water is distilled. If 0.1 ml of additional water is distilled in 15 minutes, repeat this step.

vii. If water remains in the condenser after distillation is completed, wash it down with toluene.

viii. Let receiving tube come to room temperature and read volume of water in lower layer of receiver and record volume (V) to nearest 0.01 ml.

Calculation

Percent Total Dry Matter

$$\% \text{ Total DM} = 1 - \frac{V}{W_1} \times 100$$

Where V = Volume of water in ml

W_1 = weight of sample in grams

Calculation: Percent Total Moisture:

% Total Moisture = 100 - % Total DM

Precautions

- To determine the DM of a feed accurately, the sample collected must be representative of the feed.
- To be accurate, distilled water should be analyzed for volatile acids and alcohol that can co-distill with water.
- Toluene dry matter is approximately equivalent to partial drying which leaves 3 to 5% moisture in the sample (which is roughly equivalent to volatile acids lost during drying) and is adequate for dry matter adjustment of crude protein and fiber.
- Toluene is flammable. Observe proper precautions for flammable solvents.
- Avoid inhaling vapours.
- Avoid skin contact.

Nitrogen Determination by Kjeldahl method

Protein is an important nutrient supplied by forages. When a laboratory uses wet chemistry, crude protein will most likely be measured by the standard Kjeldahl procedure. This measures total nitrogen which is then multiplied by 6.25 to arrive at the crude protein value for the forage.

Principle

Organic nitrogen when digested with boiling concentrated H_2SO_4 in presence of a catalyst (K_2SO_4 and $CuSO_4$ in the ratio of 9:1) is converted into ammonium sulphate. Ammonia is liberated by making the solution alkaline, is distilled in to a known volume of boric acid solution as ammonium borate. The ammonia collected in boric acid solution is titrated with standard H_2SO_4 or HCl solution. Then percentage nitrogen is multiplied by factor 6.25 to obtain crude protein content of the sample.

The reactions are as follows.

Organic nitrogen (in biological sample)

$$+ H_2SO_4 \rightarrow NH_3 + CO_2 + SO_4 + NO_2$$

$$(NH_4)_2 SO_4 + 2NaOH \rightarrow 2NH_4OH + Na_2 SO_4$$

$$NH_4OH + H_3BO_3 \rightarrow NH_4B (OH)_4$$

$$2NH_4B (OH)_4 + H_2SO_4 \rightarrow (NH_4)_2 SO_4 + 2H_3BO_3 + 2H_2O$$

Apparatus

- Kjeldahl Flask
- Weighing Balance
- Digestion Unit/Set
- Measuring Cylinder
- Volumetric Flask
- Funnel
- Pipette
- Microkjeldahl Steam Distillation apparatus
- Conical Flask
- Burette

Reagents

i. Sulfuric acid, concentrated, 95-98%

ii. ***Digestion mixture:*** It is the mixture of $CuSO_4$ and Na_2SO_4/ K_2SO_4 at ratio of 1:9. It acts as catalyst during digestion of feed sample for estimation of crude protein.

iii. ***2% boric acid solution (Tashiro's reagent):*** 2 gram of boric acid + 2 ml methyl red indicator + 1 ml bromocresol green + distilled water upto 100 ml. It is used to trap ammonia during distillation of sample for estimation of protein.

iv. 45% NaOH

v. N/10 H_2SO_4

vi. Mixed indicator

Procedure

The method consists of three basic steps:

1. Digestion
2. Distillation
3. Titration

Digestion

This step consists of digestion of the sample in sulfuric acid with a catalyst, which results in conversion of nitrogen to ammonia;

i. Weigh suitable quantity of sample and transfer it into a clean and dry Kjeldahl flask.

ii. Add 2-3 g digestion mixture and mix it gently.

iii. Add 30 ml H_2SO_4 and heat it at least for 10-15 minutes in inclined position. After this boil it vigorously till it becomes greenish in colour.

iv. After complete digestion allow it to cool.

v. Add about 5- 10 ml of distilled water to the Kjeldahl flask and shake properly to dissolve digested sample and transfer it into a 100 ml volumetric flask.

vi. Wash the Kjeldahl flask several times with little amount of distilled water till it becomes acid-free (test acid-free with blue litmus paper) and transfer the entire digested sample to the volumetric flask.

vii. Then make the volume upto 100 ml by adding distilled water and mark it as aliquot.

Distillation

This step consists of distillation of the ammonia into a trapping solution:

i. In a conical flask take 10 ml 2% boric acid solution, add 1-2 drops of indicator and place the conical flask below the condenser of Micro Kjeldahl Steam distillation apparatus.

ii. Now take 10 ml of aliquot into the Micro Kjeldahl Steam distillation apparatus and 15-20 ml of 40-45% NaOH to make the aliquot alkaline. Put stopper immediately.

iii. Allow distillation for 10-15 minutes. During distillation ammonia will release from the sample and trapped in 2% boric acid solution as ammonium borate in conical flask (greenish colour).

Titration

This step consists of quantification of the ammonia by titration with a standard solution.

i. Remove the conical flask after washing the tip of the condenser with distilled water into conical flask.

ii. Titrate the contents of flask against N/10 H_2SO_4 till pink colour arises.

iii. A blank may also run using all the reagents without sample to make correction of nitrogen present in the reagents used.

Calculation

1 ml. of N/10 H_2SO_4 = 0.0014 g of nitrogen

$$\% \text{ of N} = \frac{\text{ml. of N/10 } H_2SO_4 \text{ used in titration} \times 0.0014 \times \text{total vol. of aliquot} \times 100}{\text{Vol. of a liquid taken for distillation} \times \text{wt. of sample (g)}}$$

%of crude protein (CP) = % Nitrogen x 6.25

Note

i. The following table may be consulted for calculating the amount of nitrogen neutralized from NH_3 by acids of different normality.

Normality of H_2SO_4/HCl		N (g) Equivalent/ ml acid
1.	N or 1 N	0.014
2.	N/7	0.002
3.	N/70	0.0002
4.	N/10 or 0.1N	0.0014
5.	N/100 or 0.01N	0.00014

ii. K_2SO_4 is used to raise the boiling point for oxidation of material. $CuSO_4$ is added as a catalyst to speed up the reaction.

Precaution

i. All joints of distillation apparatus should be leak proof.

ii. Add few glass beads in the digestion flask to prevent bumping during digestion.

iii. No carbon particle should be present in aliquot. Presence of it indicates incomplete digestion of sample.

iv. Tip of the condenser of the Microkjeldahl Steam Distillation Apparatus must be dipped in boric acid solution to prevent loss of ammonia during distillation.

v. Cold water should run continuously through the condenser.

vi. Liquor ammonia or any ammonia emitting compound should not be kept at the place of sulphuric acid digestion, distillation and titration, particularly calcium should not be titrated in the same room.

Estimation of total ash content

Principle

A dried, ground sample is ignited in a furnace at 600°C to oxidize all organic matter. Ash is determined by weighing the resulting inorganic residue.

Apparatus

- Silica crucible
- Weighing balance
- Heater
- Metal tong
- Asbestos sheet
- Muffle furnace
- Desiccator

Procedure

- Weigh a clean and dry silica crucible (W_1).
- Place moisture-free sample in silica crucible and record its weight (W_2).
- Decarbonise the sample on a heater to make it smoke-free.
- With the help of metal tong, place the crucible in a muffle furnace and ignite it at 600°C for 2 hrs.
- Cool the muffle furnace at 150°C and remove the silica crucible

from the muffle furnace and place it on an asbestos sheet for cooling.

- Place the silica crucible in desiccator for further cooling and then take the weight of silica crucible with total ash (W_3) .

Observations

Weight of empty silica crucible = W_1 g

Weight of silica crucible with sample = W_2 g

$\therefore$ Weight of sample = $(W_2 - W_1)$ g = A

Weight of silica crucible with ash = W_3 g

$\therefore$ Weight of ash = $(W_3 - W_1)$ g = B

$$\therefore \text{Percentage of ash} = \frac{(W_3 - W_1)}{(W_2 - W_1)} \times 100 \quad \text{or} \quad B/A \times 100$$

or Pecrentage of organic matter = 100 - percentage of total ash.

Precautions

- Switch off the fans and put on the exhaust while decarbonising (smoking) the sample.
- Transfer decarbonised sample into muffle by carrying in a closed container to avoid loss of sample due to air movement.
- Weigh cooled ash sample as quickly as possible since it is highly hygroscopic.
- Though silica basin can with stand reasonably high temperatures, monitor to maintain temperature of muffle furnace between 550 - 600°C to avoid breakage.
- If more samples are ashed at a time, note the serial no of silica basins placed inside the muffle as there is a probability of loss of identification during ashing.

Estimation of acid insoluble ash (AIA)

Principle

The inorganic portion of total ash which is not soluble in HCl is known as acid insoluble ash.

Apparatus

- Silica crucible
- Weighing balance
- Muffle furnace
- Metal tong
- Asbestos sheet
- Desiccator
- Heater
- Beaker
- Funnel filter paper
- Volumetric flask
- Pipette
- Glass rod

Reagents

- Distilled water
- Conc. HCl
- Conc. HNO_3

Procedure

- Add 10 ml distilled water to the silica crucible containing total ash.
- Add 10 ml HCl and few drops of conc. HNO_3 and stir it with glass rod.
- Heat the crucible for 2 - 3 minutes and allow the crucible to cool.
- Then filter the ash solution through Whatman filter paper No. 40/42 into a volumetric flask. Then wash the silica basin with hot distilled water several times and transfer the entire insoluble residue to the filter paper. Then wash the insoluble residue with hot distilled water till it becomes free from acid (test acid-free by blue litmus paper).
- Acid soluble ash is collected in volumetric flask and acid insoluble ash remains on filter paper.

- Transfer the filter paper containing the acid insoluble ash to the same silica basin and place it on a heater to burn the filter paper. Then place the silica crucible in muffle furnace and ignite it at 600°C for 1-2 hour.
- Remove the silica crucible from the muffle furnace and place it on an asbestos sheet for cooling. Then place it in a desiccator for further cooling and record its weight.

Observations

Weight of empty silica crucible = W_1 g.

Weight of silica crucible with feed sample = W_2 g.

∴ Weight of sample = $(W_2 - W_1)$ g.

Weight of silica crucible with acid insoluble ash = W_3 g.

∴ Weight of acid insoluble ash = $(W_3 - W_1)$ g.

$$\therefore \text{Acid insoluble ash } (\%) = \frac{(W_3 - W_1)}{(W_2 - W_1)} \times 100$$

Estimation of Calcium

Principle

Calcium is present in the form of carbonate and oxide in the ash. When ash is treated with Hcl, salts are converted into calcium chloride, which in turn is precipitated as calcium oxalate by the addition of ammonium oxalate. When the precipitate is treated with sulphuric acid H_2SO_4 all the precipitate will dissolve and there will be production of calcium sulphate. Then the solution containing calcium sulphate is titrated against N/10 $KMnO_4$ to obtain calcium content of the sample.

The reaction takes place as follows.

$$\text{Ash} \rightarrow Ca\,CO_3 + 2\,HCl \rightarrow Ca\,Cl_2 + H_2O + CO_2$$

$$Ca\,Cl_2 + (NH_4)_2\,CO_4 \rightarrow CaC_2O_4 + 2NH_4\,Cl$$

↓

ppt. of Calcium Oxalate

$$Ca\,C_2\,O_4 + H_2SO_4 \rightarrow H_2C_2O_4 + CaSO_4$$

$$5H_2C_2O_4 + 2KMnO_4 + 3H_2SO_4 \rightarrow K_2SO_4 + 2MnSO_4 + 8H_2O + 10CO_2$$

Calcium is precipitated at a pH 4.0 to prevent interference by the phosphate.

Apparatus

- Beaker 250 ml
- Pipette 25,50 ml
- Burette10,25/50 ml
- Wash bottle
- Glass Rod
- Measuring cylinder 100ml
- Hot Plate
- Whatman Filter Paper No. 40
- Funnel

Reagents

- Conc. HCl
- Methyl red indicator
- Saturated solution of ammonium oxalate: Dissolve ammonium Oxalate in distilled water. Heat the solution and again add ammonium oxalate to it till ammonium oxalate precipitate in the solution i.e., upto maximum conc.
- *1:10 Ammonia Solution:* Dissolve 1 part ammonia in 10 parts of distilled water.
- *1:4 H_2SO_4:* Dissolve 1 part H_2SO_4 in 4 parts distilled water.
- *1:10 HCl Solution:* Dissolve 1 part of HCl in 10 parts of distilled water.
- *N/10 $KMnO_4$*

Procedure

i. Transfer 25 ml of aliquot (mineral extract after removals of acid insoluble ash) in the 250 ml beaker.

ii. Then dilute with an equal amount of distilled water.

iii. Add 2-3 drops of methyl red indicator.

iv. Add ammonium hydroxide solution drop wise till a brownish orange colour is obtained and few drops of HCl, so that colour of solution turns to pink (pH 2.5-4).

v. Dilute to about 100 ml with glass distilled water and bring to boiling.

vi. Now add slowly with constant stirring about 10 ml of hot solution of saturated ammonium oxalate.

vii. If red colour changes to yellow or orange, add HCl solution drop wise until colour again changes to pink.

viii. Cool the contents and keep under bell jar over night for formation of white course granules of calcium oxalate.

ix. Next Day filter the solution through Whatman filter paper No. 40.

x. Wash the beaker several times with hot distilled water and transfer all the precipitate to filter paper.

xi. Wash the precipitate with hot distilled water till it becomes oxalate free.

xii. ***Oxalate Free Test:*** *Take 5 ml of filtrate. Add 3-4 drops of 1:4 H_2SO_4. Add 1 drop N/10 $KMnO_4$. Heat to boil. If pink colour persists, then the precipitate is free from oxalate.*

xiii. Now transfer the filter paper with oxalate free precipitate from funnel to the same beaker unfold the filter paper and hang it untitled in the beaker with the support of obliquely placed glass rod below the filter paper.

xiv. Drop dilute H_2SO_4 over the fitter paper to dissolve the Ppt..The washed filter paper is transferred back to the funnel.

xv. The dissolved Ppt. in beaker is heated upto 60-70°C.

xvi. Titrate the solution against N/10 $KMnO_4$ till faint (dim) pink colour appears (faint pink colour persisting for at least 30 seconds indicates that the titration is complete)

Calculation

1ml of N/10 $KMnO_4$ = 0.00204 g of calcium.

$$\% \text{ Calcium} = \frac{\text{ml of N/10 } KmnO_4 \times 0.00204 \times \text{total vol. of mineral extract} \times 100}{\text{Vol. of a liquid taken for test} \times \text{Dry Wt. of Sample (g.)}}$$

Note

i. *To express in terms of CaO multiply %age of calcium by 1.4.*

ii. *To convert CaO to calcium multiply CaO % by 0.714*

Precaution

- Maintenance of acidic pH is necessary for precipitation of calcium oxalate.
- While stirring contents in beaker, breading of lumps of Ppt. should be avoided.
- Standard $KMnO_4$ solution be kept in an amber colour bottle at a dark place away from light.

Estimation of Phosphorus

Phosphorus in mineral extract is present in the form of

a. Disodium hydrogen phosphate (Na_2HPO_4)

b. Sodium dihydrogen phosphate (NaH_2PO_4)

c. Sodium Phosphate (Na_2PO_4)

d. Calcium dihydrogen phosphate Ca $(H_2PO_4)_2$

e. Calcium hydrogen phosphate ($CaHPO_4$)

Principle

When a solution containing phosphorus is treated with ammonium molybdate in nitric acid, phosphorus is precipitated as ammonium phosphomolybdate (yellow Ppt.). This precipitate is dissolved in little volume of known quantity of N/10 NaOH and titrated against N/10 HCl/H_2SO_4 or HNO_3 to obtain phosphorus content of sample.

The reactions are as follows:

$$Na_2HPO_4 + 12(NH_4)_2\ MoO_4 + 23HNO_3$$

$$\downarrow$$

$$(NH_4)_3\ PO_4.\ 12MoO_3 + 2NaNO_3 + 21NH_4\ NO_3 + 12H_2O$$

(ammonium phosphomolybdate)

$$\downarrow$$

$$(NH_4)_3\ PO_4.\ 12MoO_3 + 23NaOH$$

$$\downarrow$$

$$11Na_2\ MoO_4 + Na\ (NH_4)\ HPO4 + (NH_4)\ MoO_4 + 11H_4O$$

Apparatus

- Beaker 250 ml.
- Pipette
- Burette
- Glass rod
- Funnel
- Whatman filter paper No. 40 or 42

Reagents

- ***20% Ammonium Molybdate:*** 20 g Ammonium Molybdate + 40 ml distilled water + 40 ml liquid ammonia solution.
- Conc. HNO_3.
- N/10 NaOH
- N/10 HCl
- Phenolphthalein Indicator
- 3% Potassium Nitrate Solution (KNO_3)

Procedure

i. Pipette out 25 ml of aliquot (soluble ash solution) in a 250 ml beaker.

ii. In a separate beaker prepare a solution by pouring together 10 ml of 20% ammonium molybdate and10 ml of conc. HNO_3 at a time.

iii. Then quickly mix this solution with the aliquot and stir it by a glass rod. Keep the beaker overnight under a bell Jar for complete yellow precipitation of ammonium phosphomolybdate.

iv. Next day filter the solution through a Whatman filter paper No. 40 or 42.

v. Wash the precipitate with 3% potassium Nitrate till it becomes acid-free.

   a. ***Acid-Free Test:*** *Take 1 drop of N/10 NaOH. Add few drops of distilled water. Add 1 drop phenolphthalein. Add 20 drops of filtrate. If pink colour exist, then the precipitate is acid-free. Acid free test can also be performed with blue litmus paper.*

vi. Transfer the filter paper with ppt. to the same beaker, put 1 drop of phenolphthalein and add measured quantity (10 ml) of N/10 NaOH to dissolve yellow precipitate which changes into pink colour.

vii. Titrate the solution against N/10 HCl till the solution becomes colourless.

Calculation

Volume of NaOH used up = ml. of N/10 NaOH taken – ml of N/10 HCl used in titration

1 ml of N/10 NaOH = 0.000135 g of phosphorus

1 ml of N/7 NaOH = 0.0001925 g of phosphorus

$$\% \text{ of phosphorus} = \frac{\text{Vol. of N/10 NaOH} \times 0.0001347 \times \text{total vol. of a liquid} \times 100}{\text{Vol. of a liquid taken for test} \times \text{wt. of sample taken (g)}}$$

Precaution

i. Precipitate should not be washed with water due to its solubility.

ii. Content should not be heated for precipitation to avoid the formation of molybdic anhydride.

iii. If any precipitate is formed by mixing concentrated nitric acid and ammonium molybdate solution in a separate beaker, it should be discarded and only clear hot mixture of two reagents should be used for precipitation.

iv. Loss of precipitate by creeping is avoided by allowing complete decanting after each washing and transferring only small volume of content on the filter paper.

Note: *To express in terms of phosphorus pentaoxide (P_2O_5) multiply % of phosphorus by 2.29.*

Determination of ether extract (EE)

Princple

Lipids are a group of materials that are insoluble in water but soluble in ether, chloroform, and benzene. The portion of dried sample which is soluble in ether is known as ether extract, it includes triglycerides, sterols, resins, waxes, essential oils, fat soluble pigments etc. Ether extract is estimated by extracting the known quantity of sample through ether (petroleum ether or ethyl ether) or other fat solvents like benzene, hexane, chloroform etc. for 6-8 hours in Soxhlet's extraction apparatus. During extraction ether extract from the sample is collected in oil flask. The ether is then evaporated from the fat solution. The resulting residue is weighed and referred to as ether extract or crude fat.

Apparatus

- Soxhlet extraction apparatus.
- Extraction Thimble (Whatman Filter Paper No. 1)
- Oil Flask
- Water Bath/Heater with temperature controlling device.
- Balance
- Desiccator
- Hot Air Oven

Reagents

Petroleum Ether or Ethyle Ether or Acetone

Procedure

i. Weigh 1-5 g of moisture-free sample (depending upon fat content and bulk) and transfer it into an extraction thimble. Plug the top of the thimble with a thin layer of cotton.

ii. Place the thimble inside the soxhlet extractor over a cotton pad so that top of the thimble is well above the top of the siphon.

iii. Fit the soxhlet extractor with previously weighed clean and dry oil flask at the bottom and with the condenser at the top.

iv. Place the oil flask on a water bath or on a heater with temperature controlling device.

v. Pour petroleum ether or ethyl ether through the condenser. The amount of ether should be equal to 1½ siphon.

vi. Turn on water to condensers.

vii. Heat it at 60°C and allow the extraction to proceed for 8 hours (at a condensation rate of 5–6 drops per second).

viii. After extraction is over remove the thimble with fat–free sample from the soxhlet extractor.

ix. Assemble the apparatus again and heat it to recover all the ether from the oil flask.

x. Disconnect the oil flask from the Soxhlet extractor and place it in an oven for drying at 100°C for 1 hour. The oil flask now contains only the ether extract. Place the oil flask in a desiccator and weigh.

Calculation

Percent Crude Fat/ Ether Extract (DM basis)

$$\text{Ether extract (\%)} = \frac{\text{Wt. of oil flask with ether extract} - \text{Wt. of empty oil flask X 100}}{\text{Wt. of sample (g)}}$$

Note: *Preserve ether extracted sample for crude fibre estimation.*

Precaution

- Running water should flow continuously through the condenser.
- Top of the thimble should be well above the top of the siphon.
- Condensed ether from the condenser must drop inside the thimble.
- Ether extraction should be done for 8 hours preferably to avoid error due to duration of extraction.

- Thimble and sample should not be touched with hands to avoid contamination by lipid in sweat.
- There should not be any flame at the time of ether extraction.
- Make sure all ether is evaporated from the beakers before placing them in the oven to avoid a fire or explosion.
- Both the ether and the samples must be free of moisture to avoid coextraction of water-soluble components in the sample such as carbohydrates, urea, lactic acid, glycerol, etc.
- Low temperatures are used to evaporate the ether and remove residual moisture to prevent oxidation of the fat.

Determination of crude fibre (CF)

Crude fibre includes cellulose, some part of hemicellulose and a major portion of lignin. Forages vary widely in fiber content. Fiber content is the best estimate of how digestible the forage is and how much of it the animal will eat. Well managed pasture or hay can be low in fiber and highly digestible. Late-cut hay is usually high in fiber but low in digestibility and intake.

Principle

When fat-free sample is boiled in weak acid for 30 minutes followed by boiling in weak alkali for 30 minutes, all the organic matter except crude fibre will dissolve in acid and alkali solution (1.25 %). The residue left after boiling contain crude fibre and total ash. Then the sample is ignited at 600°C for 2 hours. During ignition crude fibre will be lost from the sample and loss in weight after ignition is the amount of crude fibre present in sample.

Apparatus

- Spoutless beaker (1000 ml capacity)
- Bulb condenser
- Muslin Cloth
- Horn spatula
- Silica Crucible
- Measuring Cylinder
- Tripod Stand

- Balance
- Muffle Furnace
- Metal Tong
- Desiccator
- Hot Plate

Reagents

- 10% H_2SO_4 or 1.25% H_2SO_4
- 10% NaOH or 1.25% NaOH
- 1% HCl
- Ethyle Alcohol
- Ethyle Ether

Procedure

i. Transfer fat-free sample from thimble into a spoutless beaker which is previously marked at 200 ml.

ii. Add little amount of hot water and 25 ml of 10% H_2SO_4. Then add hot water upto 200 ml mark to make the strength of acid 1.25% and boil it for 30 minutes on a hot plate at constant volume by putting condenser on the beaker.

iii. Filter the content of beaker through muslin cloth and wash the residue with hot water till it becomes acid-free (test acid-free with litmus paper)

iv. Transfer the residue from the muslin cloth to the same beaker and add little amount of hot water and 25 ml. of 10% NaOH. Then add hot water upto 200 ml mark to make the strength of alkali 1.25% and boil it for 30 minutes on a hot plate at constant volume.

v. Filter the content of beaker through muslin cloth and wash the residue with 5–10 ml of 1 % HCl and then wash with hot water till residue becomes free from alkali (test alkali free by red litmus paper).

vi. Then wash the residue with 5 ml ethyl alcohol twice and once with 5 ml ethyle ether.

vii. Transfer the residue from the muslin cloth to silica crucible and dry it at 100°C till constant weight in a hot air oven.

viii. Record the dry weight of the residue.

ix. Ignite the residue at 600°C for 2 hours in a muffle furnace.

x. Record the loss in weight after ignition. Loss in weight after ignition is equal to the amount of crude fibre present in the sample.

Calculation

Weight of sample = W_1 g

Weight of crucible + dry residue = W_2 g

Weight of crucible + total ash = W_3 g

$\therefore$ Weight of crude fibre = $(W_2 - W_3)$ g

$$\therefore \ \% \text{ of CF} = \frac{(W_2 - W_3) \times 100}{W_1}$$

Precaution

- Boiling should be done at constant volume by placing a bulb condenser at the top of the spout less beaker.
- There should not be any particle of sample sticking to the side of the beaker above the solution.
- Muslin cloth should be free from starch and contain 18-22 threads per cm of cloth.
- Frothing during boiling may be controlled by using few drops of amyl alcohol.

Estimation of neutral detergent fibre (NDF)

Neutral detergent fiber (NDF) is an estimate of the plant's cell wall content and includes the ADF and hemicellulose. Some of the NDF is highly digestible. Forage NDF is the best indicator of how much forage a high-producing animal will eat.

Principle

When feed sample is boiled (refluxed) in neutral detergent solution for one hour, all the cell content will dissolve in neutral detergent solution. The residue left after boiling is known as cell wall content. The cell wall content is then dried at 100-105°C for overnight and weighed to determine % cell wall content present in sample. Cell content is determined by difference i.e., %cell content = 100 - % cell wall content.

Apparatus

- Spoutless beaker (1000 ml)
- Bulb condenser
- Hot plate
- Sintered glass crucible/Gooch crucible(Grade-1) of 50ml capacity
- Hot air Oven
- Vacuum pump / Filter pump
- Weighing Balance

Reagents

- Neutral detergent solution
- Decahydronaphthalene
- Acetone
- Sodium sulphite

Composition and preparation of Neutral Detergent Solution (pH 6.9-7.1)

Sodium- lauryl sulphate = 30 g

Di Sodium ethyle diamine tetraacetate (EDTA) = 18.61 g

Sodium boratedecahydrate = 6.81 g

Disodium Hydrogen phosphate = 4.56 g

Ethylene glycol = 10 ml

Distilled water upto 1000 ml

Put EDTA and sodium borate decahydrate ($Na_2B_4O_7.10H_2O$) in a large breaker (2 litres). Add distilled water, shake and heat until these are dissolved. Now add sodium lauryl sulphate and 2-ethoxy ethanol. Add disodium hydrogen phosphate (Na_2HPO_4) in a separate beaker and some distilled water and heat until it is dissolved. Mix both the solutions properly and check pH to range between 6.9 and 7.1. If made properly pH adjustment will rarely be required.

Procedure

i. Weigh 0.5-1.0 g dried sample (ground to pass through 20-30 mesh screen) and transfer it into a spoutless beaker.

ii. Add 100 ml cold (Room Temperature) neutral detergent solutions.

iii. Add 2 ml decahydronepthalene as anti-foaming reagent. Not needed if 1000 ml capacity beakers are

iv. Add 0.5 g Sodium sulphite.

v. Heat to boiling for 5-10 minutes. Reduce heat as boiling begins in order to avoid foaming. Reflex for 60 min. timed from onset of boiling.

vi. Filter the content of beaker through a previously weighed Gooch crucible with the help of filter pump.

vii. Wash the residue with hot water (90-100°C) and then wash twice with acetone.

viii. Dry the crucible with residue (NDF) at 100°C for 8 hrs.

ix. Cool in desiccator and take weight of the crucible with NDF.

x. Ash the residue for 3 hours at 550°C and weight after cooling. Report ash content as insoluble in neutral detergent.

Calculation

Weight of sample = W_1 g.

Weight of Empty crucible = W_2 g

Weight of crucible + NDF = W_3 g

Weight of NDF = $(W_3 - W_2)$ g

$$\therefore \% \text{ of NDF} = \frac{W_3 - W_2 \quad x\ 100}{W_1}$$

$\therefore$ % of cell content = 100 - %NDF

Ash insoluble in neutral detergent = wt. of crucible + Ash – wt. of empty crucible.

Estimation of acid detergent fibre (ADF)

Acid detergent fiber (ADF) is the less-digestible cellulose and lignin or "woody" fiber in the plant. ADF is an indicator of digestibility across different species of grasses and legumes. As ADF increases, digestibility decreases.

Principle

When sample is boiled in acid detergent solution for an hour all the cell content and hemicellulose will dissolve in acid detergent solution. The residue left after boiling is known as acid detergent fibre. Acid detergent fibre (ADF) is then dried and weighed to determine %ADF present in sample. Hemicellulose is determined by difference between NDF and ADF.

Apparatus

- Spoutless beaker (1000 ml)
- Bulb condenser
- Hot plate
- Sintered glass crucible/Gooch crucible (Grade-1) of 50 ml capacity
- Hot air Oven
- Vacuum pump / Filter pump
- Weighing Balance

Reagents

- Acid detergent solution
- n-hexane
- Decahydronapthalene (Decalin)
- Acetone

Acid detergent solution

Add 20 gram of Cetyl trimethyle ammonium bromide (CTAB), Technical grade in 1 N H_2SO_4 (27.7 ml of concentrated reagent grade sulfuric acid to 972.3 ml water) and make the volume to one liter.

Procedure

i. Weigh 1g dried sample and transfer it into a spout less beaker.

ii. Add 100 ml cold (Room Temperature) acid detergent solution and 2 ml. decahydronapthalene (decalin)

iii. Heat to boiling for 5-10 minutes. Reduce heat as boiling begins in order to avoid foaming. Reflex for 60 min. timed from onset of boiling.

iv. Filter the content of beaker through a previously weighed gooch crucible.

v. Wash the residue (ADF) twice with hot water and then wash with acetone until it removes no more colour.

vi. Wash with hexane. Hexane should be added while crucible still contains some acetone. Washing with hexane can be avoided if lumping is not a problem.

vii. Suck the acid detergent fibre free of hexane and dry at 100 – 105°C for 8 hours.

viii. Cool in a desiccator and take the weight of crucible with ADF.

Calculation

Weight of sample = W_1 g

Weight of Empty crucible = W_2 g

Weight of crucible + ADF = W_3 g

Weight of ADF = $(W_3 - W_2)$ g

$$\therefore \% \text{ of ADF} = \frac{W_3 - W_2 \times 100}{W_1}$$

Again NDF = ADF + hemicellulose

∴ % of Hemicellulose = % NDF - % ADF

However, this difference also includes some fibre-bound protein. But this is negligible.

Safety Precautions

i. Always add sulfuric acid to water. Wear face shield and heavy rubber gloves. If acid is splashed on skin, wash immediately with copius amounts of water.

ii. CTAB powder irritates mucous membranes, eyes and skin. Wear gloves and dust mask while handling.

iii. Acetone is highly flammable. Do not let vapors accumulate in work area. Use effective fume removal device. Also avoid inhaling or contact with skin. Make sure all traces of acetone have evaporated from the crucibles containing fiber residue before placing in the drying oven.

Estimation of lactic acid in silage

Lactic acid is the strongest and most abundant acid produced during an ideal fermentation. Moderate levels of lactic acid are an indication of good fermentation. The lactic acid content of good maize and alfalfa silage ranges from 4 to 6% and from 3 to 8%, respectively on a dry matter (DM) basis. In high moisture maize, lactic acid is normally found in the 1 - 3% range.

Principle

Lactic acid in silage is converted into acetaldehyde by treatment with concentrated sulfuric acid, and the acetaldehyde is determined by its color reaction with p-hydroxydiphenyl in the presence of cupric ions. The color is read in a photoelectric calorimeter with a filter having a peak transmission at 560 mμ (Baker and Summerson, 1941).

Reagents

- ***Copper sulphate solution* (20%):** Dissolve 200 g of $CuSO_4.5H_2O$ in 500 ml distilled water and make the voloume to one litre.
- ***Copper sulphate solution* (4 %):** Dissolve 40 g of $CuSO_4.5H_2O$ in 500 ml distilled water and make the volume to one litre.
- **Sulfuric acid, concentrated**, sp. gr. 1.84.
- ***Solid calcium hydroxide***
- ***Sodium hydroxide (5%)*** : Dissolve 5 gram NaOH in 90 ml distilled water and make up volume upto 100 ml.
- ***p-hydroxydiphenyl reagent***: Take 1.5 gram of p-hydroxydiphenyl in a 100 ml volumetric flask. Add 10 ml of 5% NaOH and 10 ml distilled water. Warm it with constant stirring to dissolve. Make the volume to 100 ml. Store it in amber colour bottle.
- ***Stock lactic acid solution***: Take 0.1065 g lithium lactate in 100

ml volumetric flask and dissolve in 50 ml distilled water. Add 0.1 ml concentrated sulphuric acid and make up the volume to 100 ml with distilled water. The solution contains 1 mg lactic acid per ml and the solution is stable for a long period in refrigerator.

- ***Working standard lactic acid solution***: Dilute 1 ml of stock lactic acid solution to 100 ml with distilled water. It contains 0.01 mg lactic acid per ml.

Processing of the silage samples

Weigh 10 grams of freshly taken silage samples in a beaker. Add about 50 ml distilled water, transfer the contents of beaker in homogenizer and macerate the content for 10 minutes. Filter the content of homogenizer in a 100 ml capacity volumetric flask using fine glass wool. Make up the volume with water upto mark. Measure 1 ml of silage extract in 10 ml graduated centrifuse tube in duplicate. Add 7 ml distilled water, 1 ml of 10 percent sodium tugstate, 1 ml of two thirds normal sulphuric acid. Shake the centrifuse tube through thoroughly and make up the volume upto 10 ml. the dilution will be 1:10. Wait for about 15 minutes and then centrifuse the contents of tube at 3000 rpm for 15 minutes. Transfer the supernatant in a separate centrifuse tube.

Procedure

1. Add 1 ml of supernatant of silage sample in a centrifuse tube. Add 1 ml of 20% copper sulphate and make the volume to 10 ml.
2. Add 1 gm. of powdered calcium hydroxide and shake vigerously to make the mixture homogenous.
3. Leave the tubes for 90 minutes at room temperature, with occasional shaking.
4. Centrifuge at 3000 rpm for 10 min.
5. Take 1 ml supernatant in a test tube in duplicate.
6. Add 0.05 ml of 4% copper sulphate.
7. Add 6 ml concentrated sulphuric acid drop by drop with continuous shaking.
8. Keep the tubes in boiling water bath for 5 minutes.

9. Cool the tubes at room temperature.
10. Add 0.1 ml p-hydroxydiphenyl reagent drop by drop. Mix the contents immediately and vigerously.
11. Incubate the tubes at 30°C for 30 min with periodic shaking.
12. Keep the tubes in boiling water bath for 90 sec. Remove the tubes and cool to room temperature.
13. To plot the standard curve, prepare the standard tubes in duplicate as follows:

Tube No.	Blank	I	II	III	IV	V
Distilled water (ml)	1.0	0.9	0.8	0.6	0.4	0.2
Standard lactic acid solution (ml)	0.0	0.1	0.2	0.4	0.6	0.8
Lactic acid (μg) in the tubes	0.0	1.0	2.0	4.0	6.0	8.0

14. Read absorbance "A" of the tubes at 560nm against reagent blank.
15. Find out the concentration of sample on standard curve and multiply by 10 (dilution) to calculate μg lactic acid/ml in silage sample.

References

AOAC. 2000. *Official Methods of Analysis*, 17th ed. Association of Official Analytical Chemists, Washington, DC.

Barker, B. and. Summerson, W.H. 1941. The colorimetric determination of lactic acid in biological material. *J. Biol. Chem.*, 138:535-554.

Dewar, W.A. and P. McDonald. 1961. Determination of dry matter in silage by distillation with toluene. *J. Sci. Agric.* 12:790-795.

Fenner, H. 1984. Methods for determining fermentation products in acid preserved feeds and forages. Massachusetts Agricultural Experiment Station. *Res. Bull.* 691. pp. 14

Goering, H.K. and Van Soest, P.J. 1970. Forage fiber analysis (apparatus, reagents, procedures, and some applications). USDA Agricultural Research Service. Handbook number 379.

Goering, H.K., Gordon, C.H., Hemken, R.W., Waldo, D.R., Van Soest, P.J. and. Smith, L.W. 1972. Analytical estimates of nitrogen digestibility in heat damaged forages. *J. Dairy Sci.* 55(9): 1275-1280.

Mertens, D.R. 1992. Critical conditions in determining detergent fiber. Proceedings of NFTA Forage Analysis Workshop. Denver, pp C1-C8.

Van Soest, P.J, J.B. Robertson, and Lewis, B.A. 1991. Methods for dietary fiber, neutral detergent fiber and non-starch polysaccharides in relation to animal nutrition. *J. Dairy Sci.* 74:3583-3597.

Subject Index